Bat Roosts in Rock

Frontispiece Anyone can find horseshoe bats in rock; it takes real determination and skill to find a whiskered bat *Myotis mystacinus* behind a sandstone flake, and retrieve a dropping to get the identification confirmed by DNA analysis. © Rob Bell

Bat Roosts in Rock

A Guide to Identification and Assessment
for Climbers, Cavers & Ecology Professionals

Bat Rock Habitat Key

Pelagic Publishing | www.pelagicpublishing.com

First published in 2021 by
Pelagic Publishing
PO Box 874
Exeter
EX3 9BR, UK

www.pelagicpublishing.com

*Bat Roosts in Rock: A Guide to Identification
and Assessment for Climbers, Cavers & Ecology Professionals*

© Henry Andrews 2021

British Library Cataloguing in Publication Data
A catalogue record for this book is available from the British Library

https://doi.org/10.53061/RKRW8270

ISBN 978-1-78427-261-6 *Paperback*
ISBN 978-1-78427-262-3 *ePub*
ISBN 978-1-78427-263-0 *PDF*

Cover images:

Top (main) image: A common pipistrelle *Pipistrellus pipistrellus*
recorded behind a gritstone flake by Rob Bell and his comrades from
South Yorkshire Bat Group. © Rob Bell

Bottom left: A Natterer's bat *Myotis nattereri* roosting high in a limestone
alicorn. © Henry Andrews

Bottom middle: Sam Dyer going the extra mile to record a Welsh serotine
Eptesicus serotinus roost behind a limestone flake 20 m up a worked-out
quarry face. © Sam Dyer

Bottom right: A Welsh brown long-eared bat *Plecotus auritus* recorded in
a sandstone crack by Hal Starkie and his roost-finding posse. © Hal Starkie

Printed and bound in Wales by Gomer Press Ltd.

MIX
Paper from
responsible sources
FSC® C114687

Contents

Preface

Much of the information that has historically been published about bats in rock in the British Isles was not formatted with practical application in mind. Regardless, in the early 2020s our basic knowledge of exactly how bats exploit rock remains patchy.

In the case of rock faces, historically we have had: Sam Dyer's work with serotines *Eptesicus serotinus*; the results of a study by Mike Shewring; two photographic accounts of individual barbastelle *Barbastella barbastellus* and brown long-eared bat *Plecotus auritus* from Jean Matthews and Nathalie Cossa respectively, but aside from a lot of smoke and mirrors, that really was it.

There has been Scandinavian work published on scree, but in the British Isles there have been just vague musings and little else.

The accounts of subterranean roosts that were published in the British Isles and Europe in the 1960s and 1970s are comprehensive when collated, and both minutely detailed and jaw-dropping in their scope. But these accounts appear not to have been widely read. Meanwhile, some accounts that have been widely cited might have unintentionally been misinterpreted.

Even now, some accounts that might improve our understanding remain out of circulation, and could not therefore be collated into this book. In particular, we would like to see:

» Baagoe, H., Degn, H. & Nielsen, P. 1988. Departure dynamics of *Myotis daubentonii* (Chiroptera) leaving a large hibernaculum. *Vidensk. Meddr. Dansk naturh. Foren.* 147: 7–24.

» Jurczyszyn, M. 1998. The dynamics of *Myotis nattereri* and *M. daubentonii* (Chiroptera) observed during hibernation season as an artefact in some type of hibernacula. *Myotis* 36: 86–91.

In combination, the historical British accounts, together with both historical and more modern Scandinavian and European accounts, offer the potential for the definition of a framework for data collection. This framework could be used to identify the environmental niche of our resident species, and that data might thereon be used to: **a)** identify existing potentially important rock habitats; **b)** inform enhancement projects; **c)** inform designs for entirely new 'artificial' habitats in situations where they must be built 'from scratch'; and, **d)** define environmental trigger thresholds for monitoring programmes that can be adopted as a definition of success.

In 2021 we have the technological gadgetry to gather the data needed to understand why different bat species occupy different sites in different numbers at different times. Once collected this data might be collated on one database and made accessible to everyone (see www.batrockhabitatkey.co.uk).

We finally have the means to communicate this information across the length and breadth of the British Isles and the wider network of ecologists around the world in Technicolor detail, and has long since been identified:

Full disclosure of data is an elementary precaution against both honest mistakes and charlatans.

Peters, R. 1991. *A Critique for Ecology*. Cambridge University Press.

It is likely that the lion's share of the work will be done by amateur naturalists, but nevertheless it will ensure professional ecologists have an evidence base to draw from. Ultimately, it is hoped that the book will benefit the bats.

Citing, credits and acknowledgements

Citing

Where reference is made to any part of the content of this book, please cite as 'BRHK 2021' giving the full reference as:

> » BRHK 2021. *Bat Roosts in Rock – A Guide to Identification and Assessment for Climbers, Cavers & Ecology Professionals.* **Pelagic Publishing, Exeter.**

Credits

Text

Chapter credits are as follows:

> » **Chapter 1** – Henry Andrews
> » **Chapter 2** – Henry Andrews & Steven Hopkins
> » **Chapter 3** – Henry Andrews, Robert Bell, James McGill & Hal Starkie
> » **Chapter 4** – Henry Andrews & James McGill
> » **Chapter 5** – Henry Andrews
> » **Chapter 6** – Henry Andrews

Henry Andrews is the founder of both the *Bat Tree Habitat Key* and *Bat Rock Habitat Key* projects. He has a particular interest in the use of structure-based predictive frameworks and data-driven analysis in the definition of hypotheses against which to compare the results of surveys.

Steven Hopkins is an engineering geologist with 20 years of experience and has worked in the quarrying industry since 2006. He is chartered with the Geological Society of London and is a Fellow of the Institute of Quarrying. His interest lies in how we use and engineer the natural geology for the benefit of society.

Robert Bell is a long-term member of South Yorkshire Bat Group with a keen interest in bat conservation. He is a partner in Middleton Bell Ecology, a small ecological consultancy based in Barnsley.

Dr James McGill is an entomologist, ornithologist, and botanist.

Hal Starkie is a bat ecologist who undertakes multiple research projects within the south Wales region, including studies of woodland *Myotis* bats, Nathusius pipistrelles, bat roosts within rock faces and swarming bats. He is involved in many aspects of bat work, and acts as records officer for Cardiff Bat Group, as well as having been a registered bat carer since 2014. Hal has also worked with bats across the world including work in Europe and Africa.

Photographs

BRHK is grateful to the naturalists who have provided the individual photographic accounts we have used as illustrations, as follows:

Chapter 3

Jean Matthews for the photographs of the barbastelle *Barbastella barbastellus* roost in a break; **Hal Starkie** for photographs of the serotine *Eptesicus serotinus* roosts in a crack and bidoigt, the Daubenton's bat *Myotis daubentonii* roost in a break, and the brown long-eared bat *Plecotus auritus* roosts in a crack and behind a flake; **Sam Dyer** for the photographs of the serotine *Eptesicus serotinus* roost behind a flake; **Robert Bell** for photographs of the whiskered bat *Myotis mystacinus* roosts in a crack and behind a flake, the Natterer's bat *Myotis nattereri* roosts in a crack and behind a flake, the common pipistrelle *Pipistrellus pipistrellus* roosts in a break and behind a flake, and the soprano pipistrelle *Pipistrellus pygmaeus* roost in a crack; and, **Ian Wright** for the photographs of the common pipistrelle roost in a crack.

Chapter 5

Stuart Spray Wildlife for photographs of the complexity and scale of mines, the sloping silt-covered shelf which gives way to a deep and cold sump, the subterranean lake in a mine system, the mining hall, and the simple and complex caverns; **Paul Bowyer** for the photograph of tissue *Triphosa dubitata*; **Colin Morris** for the photographs of whiskered bats roosting in pleats; and, **Rich Flight** for the images of a vertical pothole entrance, and various bats roosting in pleats, cracks, breaks, flowstone, and alicorns.

Acknowledgements

Henry Andrews – My thanks to George Bemment for showing me Oak Cave and attempting to inspire me to take an interest in cave-roosting ecology over a decade ago; see, I was listening.

Thanks also: to Chris Barrington for thinking to send me Coward's 1907 paper, which led to a load of other interesting stuff; to Geoff Billington for supplying coastal cave data and without whom there would be no Chapter 4; to David Goodman, Graham Warrington, Alan Walters, Den Williams and Shaun Denny for all the work they have done and continue to do for the bats of Garth Iron Mine; to Steven Hopkins for adding the credibility factor to Chapter 2; to Erik Korsten for providing a wealth of white literature; to Rune Sørås (Norwegian University of Science and Technology) and Robert Schorr (Climbers for Bat Conservation) for providing information, useful discussion and enthusiasm; to Stuart Spray and Dan Bell for dragging my weedy carcass into some breath-taking caverns which I otherwise would never have seen, and for gathering data and taking photographs that no one would have otherwise seen; and, thanks to Andy Watson for showing me the hidden entrance to the weirdest rock landform I have seen so far and changing my perspective of just where bats can hide.

Thanks to Richard Crompton, Jon Russ and Stuart Spray for donating dataloggers so that the project could keep going when funds were desperately short – see, I said I would keep you all anonymous and I lied… Hell no one will read the book far less look at this bit. But we know… we know…

Thanks to Louis Pearson for accompanying me on some dubious missions without asking those little awkward questions that a lesser man might have asked, and for staying calm when things inevitably went VERY 'sideways'; to Dr James McGill for going beyond the call of duty to root out hard-to-get literature, and for all the invertebrate accounts; to Abigail Smart and Heather Gardiner for assisting with all the tabulation, data entry, and proof-reading (for sense but not removing the character from the syntax).

Thanks to Theresa Radcliffe, Erin Andrews, Eugene Andrews, Isobelle Andrews and Tristan Andrews for keeping ever so quiet while I fought my way through pulling the project together, even though we were all on lockdown.

Thanks to David Hawkins (Production Editor, Pelagic Publishing) for his dogged persistence in refusing to accept blurry figures and his endless patience, as well as BBR Design who laid out all the complex material in a logical order; your combined efforts have resulted in an 'illuminated' manuscript and I am so very pleased with the finished book.

Finally, thanks to all the BRHK collaborators who have gone out and field-tested the recording criteria to provide an initial foundation of data, contributed to the chapters and reviewed the narrative for accessibility, consistency and sense.

Robert Bell – Particular thanks should go to my wife Samantha Bell, who has always been willing to turn a blind eye as I search for bats during my free time, following a week doing a similar thing for pay. Thanks also to Rose Bell, who witnessed the finding of a crag roost on her first trip outdoors and to Seren Bell who shows a promising level of enthusiasm towards the species group.

I find the bat community a constant source of inspiration and I am lucky to live close to plenty of keen, open and enthusiastic bat workers. In particular Greg Slack, Benjamin McLean, Amanda Murphy and Peter Middleton rarely turn down the opportunity to go out on a whim in order to find something new. Thanks also to Tina Wiffen, Ian Bond and Lisa Worledge for providing a ready outlet for articles written about local sites and to Professor John Altringham for providing much of the early inspiration.

Finally, thanks to Henry Andrews for inviting me to input into this project and providing some much-needed structure to organise my crag roost findings.

Hal Starkie – Thanks to Jess Dangerfield who has been equally involved in surveying for bats within the rock faces with me and all the volunteers of Vale of Glamorgan and Bridgend Bat Group who have given up their free time, and braved the cold and damp weather, to assist with surveying the rock faces. And thanks to Natural Resources Wales for granting us access to survey in Ogmore Forest.

CHAPTER 1

Rationale

Henry Andrews

In this chapter	
Preamble	The motivation for producing this book
The BRHK project objectives	The four objectives
Progression	Working from the outside in: from the rocks themselves, through rock faces, onto loose rock and down into subterranean situations. How the book is structured into five subjects and how the subjects are laid out in the chapters. Where the legislation that protects bats and their roosts fits into it all
Approach	Scoping-in to collate what has been confidently proved, recorded and thoroughly described, as well as what has been alluded to in reports, in order to see where information is complete and certain, or incomplete, uncertain or entirely lacking. The approach to the environment on three levels: macro, meso and micro. The approach to the bat species ecology. An initial summary of surrogate species for use in the field
Format	Using the project recording forms and database framework as a framework for the landform chapters. Laying everything out to satisfy biological recording principles. Separating the players from the plebs

1.1 Preamble

The fact that rock habitats are exploited for roosting year-round by a variety of bat species has long been known; this knowledge is accepted by professional ecologists and amateur naturalists alike.

Such habitats encompass: **1)** surface situations where the bats exploit exposed rock faces on which they land and then crawl into their roost position; **2)** subsurface situations where the bats exploit loose rock upon which they land and crawl down into their roost position; and, **3)** subterranean (i.e. fully underground) situations into which they can fly and then find a roost position.

Each of these three situations can be further subdivided into landforms that have a natural origin and those that have an artificial origin.

Studies of roosting ecology in this context have been widely spaced over several decades by teams acting independently in a wide range of locations and situations, and for different reasons.

The greater proportion of this work took place before: **a)** Dr Robert Stebbings and others like him alerted the public to the need for an organised effort to protect the habitats upon which bats rely; **b)** radio-tracking allowed the bats to show the academics where they roosted; **c)** loggers facilitated the collection of environmental data in minute detail and over an entire year at a time; **d)** digital cameras allowed the capture of detailed images that could be reviewed in the field and distributed across the world in seconds; **e)** the word processor let anyone not only write an account that could be easily revised and enlarged, but also easily store and manipulate large volumes of detailed data; **f)** the Internet opened up a single line of communication that would allow anyone to search for relevant accounts using keywords, and having found them, collate them for a practical purpose, or use social media to publicise new findings as they happened; and **g)** almost all of the work took place before the professional ecologist appeared and found themselves faced with requests to survey natural and artificial rock habitats for bats, and advise on their management in that context.

Even accepting that everyone already knows that 'bats' roost in rock habitats, a single evidence-supported narrative that had collated the existing science and used the data to both define measurable attributes for the assessment of quality, and inform practical management advice, was always going to be a handy reference to have on the shelf. And that was the original aim of this book. However, the review of the available evidence identified that:

1. There has historically never been a single standardised approach to mapping and cataloguing Potential Roost Features (PRF) in surface, subsurface or subterranean rock habitats published anywhere in the world.

2. As so many of the studies were performed before technological advances provided temperature and humidity dataloggers, lux meters, endoscopes and ultrasound detectors, there are improvements that might be made in the detail of the information. And as all the equipment is publicly available, that detailed information might be collected by amateur naturalists and professional ecologists alike.

These realisations led to the thought that it might now be possible to gather the data needed to inform the enhancement of existing rock habitats that are not suitable for roosting bats to increase their value. And, that this data might also inform the design of artificial species-specific habitat features in situations where they had not been available before, such as within quarry restorations.

At the time of writing, the National Planning Policy Framework, which sets out the Government's planning policies for England and how these should be applied, highlights that one of the purposes of the planning system is to contribute to providing net gains for biodiversity. Developers may therefore seek to build in enhancements to project proposals, and one such enhancement is bat roost habitat.

A fact that is often overlooked is that bats are entirely subservient to their roost environment: they do not make the roosts upon which they rely.

Although it is a slow process, the habitat of tree-roosting bats can be created by *Homo sapiens* simply by neglect: over time trees will colonise and these will form roost features. These roost features naturally offer the environment favoured by bats without the need for human intervention. In contrast, rock habitats only form naturally over geological time and thus take far longer to develop than even an 'ancient' woodland.[1]

To deliver rock habitats quickly, they cannot therefore be left to appear on their own but must be artificially created with every characteristic offered by human invention. This appears relatively simple in principle: quarries deliver sheer faces and loose rock, and mining delivers adits and caverns. But although some of these man-made artificial landforms are adopted by bats, the greater proportion appear not to be. The question remains as to what separates a good landform from a bad one. If we knew this, we would know which might be left 'as is', which might be enhanced (and how) and which were really not worth the effort.

Only cursory and incidental consideration has been given to preserving rock faces and loose rock as bat habitat, and none of the landforms that have been preserved have been subject to any follow-up investigation to see if bats actually do use them. However, recently (and less recently) a handful of projects have involved considerable effort and expense in attempts to enhance artificial subterranean structures that were not already exploited by roosting bats. These have included railway tunnels, World War II bomb-shelters and also the small blockhouse bunkers known as pillboxes. There have also been attempts to create specific bat-friendly subterranean systems from scratch.

In some situations, despite what appears to be a superficially fantastic roost feature, the results have been unsatisfactory, i.e. no bats have adopted them, or a different bat species has adopted them and even then, in what are perceived as low numbers.

This may be as a result of there being only one publication that has ever offered any materially useful advice on the subject: the *Bat Workers' Manual* (Mitchell-Jones & McLeish 2004). This book remains a fantastic resource. Notwithstanding, the ideas offered in respect of rock habitats are a general 'one-size-fits-all'. Furthermore, the advice predates the widespread availability of technologically advanced environmental surveillance equipment and, as a consequence, lacks any measurable monitoring thresholds by which to test whether the enhancements suggested have in fact achieved the target environment favoured by the bat species for which the work was performed.

Unfortunately, the fact that 'bats' have adopted the habitat is sometimes presented as validation that enhancements have been successful, even when the bats present are not actually the species the work was intended to benefit.

Without knowing the environmental niche[2] the target species occupies when roosting, if that species does not exploit the enhanced or newly created habitat, there is uncertainty as to whether this is because the enhancements have failed to offer the correct environment.

The situation is such that even the desired result is simply good fortune – a happy accident. However, if the result appears to be a failure, the team cannot meaningfully assess whether: **a)** the situation might be due to something within their control to fix; or, **b)** the environment is correct and there is nothing more they can do beyond waiting for the species to find the new habitat and adopt it.

Furthermore, where the project achieves the initial objective the inquisitive may think beyond the scope of the project and ask additional questions, such as: **a)** is there anything within our control that might be manipulated to improve the environment further, or to increase the area of the favourable environment within the habitat and thereby increase the carrying capacity, allowing more of the population of Species 1 to occupy it?; **b)** is there anything within our control, that might be manipulated to encourage Species 2, which we know coexists with Species 1 in another site, to also adopt this site?; and, **c)** might further attempts to enhance the environment for a species that is not present have an associated cost, by making it less suitable for another bat species that is present?

The niche ranges on the environmental gradients can be used to predict individual species occupancy in a specific rock habitat feature. In addition, they would serve to act as the target ranges for enhancement action, and the trigger thresholds for remediation action in subsequent monitoring programmes. This would inform attempts to adjust the artificial habitat until it offered exactly the right environment to favour the target species.

1.2 The BRHK project objectives

The identification and description of those niche ranges is encompassed within the four objectives of the Bat Rock Habitat Key (BRHK) project, which are married with their definitions of success as follows:

OBJECTIVE 1

Objective – The production of a handbook that describes the Potential Roost Features (PRF) that are commonly encountered on, among and in natural and semi-natural surface, subsurface and subterranean rock habitats. The purpose of the handbook will be to inform targeted searches for PRF.

How the objective will be achieved – The production of the handbook will be achieved as follows:

» *Step 1* – Published descriptions of surface, subsurface and subterranean rock habitats will be read to identify and name the broad rock landforms, the rocks they are made of, their formation and structure. This information will be collated into an identification guide.

» *Step 2* – Visiting surface, subsurface and subterranean rock habitats to identify and describe all obvious features and environmental situations commonly encountered in and on them.

» *Step 3* – The production of an agreed set of technical terms and language to describe the appearance and attributes of the habitats and PRF so that climbers, cavers and ecologists can communicate.

Definition of success – The definition of success will be a comprehensive set of detailed accounts, comprising: location, landform, rock, formation, structure, and environment, with all the habitats and PRF accompanied by illustrative photographs and figures.

OBJECTIVE 2

Objective – The collation of published descriptions of bat roosting ecology on, among and in natural and semi-natural rock habitats.

How the objective will be achieved – The review will be achieved as follows:

» *Step 1* – Published accounts will be obtained, read and catalogued.

» *Step 2* – Information that has a practical application will be collated and written as a helpful narrative.

» *Step 3* – The narrative will be presented in a format suitable for practical application with illustrative photographs and figures.

Definition of success – The definition of success will be a single overarching document that includes as a minimum: **a)** a summary of what is confidently known about roosting ecology for each bat species, and what is not known; **b)** the identification of environmental variables that influence bat roost presence; **c)** a summary of pre-existing recording methods and criteria that might be adopted for this project; **d)** the identification of equipment and recording values that have been successfully applied across multiple studies, and might therefore be adopted for this project; and, **e)** the identification of new equipment and recording methods that might be used to further investigate roosting ecology.

OBJECTIVE 3

Objective – The production and maintenance of an online open-access database that will serve to fill gaps in knowledge and have a practical application for: **a)** the prediction of which PRF are likely to occur in a specific landform of a specific rock; **b)** the prediction of the potential for a specific PRF in a specific rock habitat to be occupied by a specific bat species; and, **c)** the creation of entirely artificial sites.

How the objective will be achieved – The database will be achieved as follows:

» *Step 1* – Recording criteria will be defined for surface, subsurface and subterranean habitats and the PRF they offer.

» *Step 2* – A database framework will be devised in Excel format.

» *Step 3* – The recording criteria will be field-tested and the data input into the Excel framework to ensure the overall database functions.

» *Step 4* – The initial databases will be offered online as an open-access resource.

» *Step 5* – Recording forms and a facility for the submission of data will be offered online.

Definition of success – The provision of a functional online open-access database that may be accessed and contributed to by anyone.

OBJECTIVE 4

Objective – Securing project funding.

How the objective will be achieved – The funding will be gained through the production and sale of a single overarching handbook for: **a)** assessing the importance of natural and semi-natural rock habitats for bats, including the identification of thresholds that will determine the suitability of a specific feature/situation to hold a specific bat species; and, **b)** artificially creating rock habitats for specific bat species.

The manual will be achieved through: **a)** securing Objectives 1, 2 and 3; **b)** the project co-ordinator(s) committing to writing the manual free of charge; and, **c)** finding a publisher.

Definition of success – The definition of success will be a complete field-guide-sized handbook published and distributed for sale.

Achieving each of the individual objectives is the subject matter of this book, and this book is the introduction to the project.

The book is designed to complement Chapter 11 of the *Bat Workers' Manual* (Mitchell-Jones & McLeish 2004) and should be read in tandem with it.

1.3 Progression

The book works from the outside in, starting with a summary review of the rocks in which bats have been recorded and reported roosting, then progressing from the surface of escarpments, crags, tors, gorges, boulders and blocks, cuttings and quarries, onto the subsurface screes, talus, blockfields and then finally into the subterranean environment of solution caves, sea caves, mines and railway tunnels.

1.3.1 Book and chapter layout

The book is broadly divided into five subjects, as follows: **1)** the rock itself; **2)** rock face landforms; **3)** loose rock landforms; **4)** subterranean landforms; and, **5)** some advice on surveying in the context of rock landforms.

Rock gets a single dedicated chapter. Thereafter the three landform categories each have individual chapters that explain how to characterise them using predefined recording criteria. Finally, there is a chapter dedicated to surveying the habitat for the bats. The chapters that characterise the habitat that the landforms offer use the survey recording form as a framework. This puts all the pertinent information into a practical context. The chapter that describes the surveying is designed to work in tandem with the *Bat Workers' Manual* (Mitchell-Jones & McLeish 2004), but offers advice on how to avoid irritation and frustration rather than a set of rules.

However, before we rush out into the hills and begin poking about, this is a convenient point to identify that bats and their roosts are legally protected in the UK, and to make reference to legislation and survey licensing.

1.3.2 Where legislation and licensing fits into all this

In the UK and the European Union (EU) all bat species and their roosts are legally protected under UK and EU law. There are currently two legislative mechanisms in force, comprising:

1. The *Wildlife & Countryside Act 1981 (& as amended)*;
2. The *Conservation of Habitats and Species Regulations 2017*.

The two pieces of legislation are set out in abridged form below.

Wildlife & Countryside Act 1981

All bat species are listed under Schedule 5 of the *Wildlife & Countryside Act 1981* and receive legal protection under Part 1, Section 9, Subsection (4)(b & c) which states:

> Subject to the provisions of this Part, a person is guilty of an offence if intentionally or recklessly —

> (b) he disturbs any such animal while it is occupying a structure or place which it uses for shelter or protection; or
> (c) he obstructs access to any structure or place which any such animal uses for shelter or protection.

Conservation of Habitats and Species Regulations 2017

All bat species are listed under Schedule 2 of the *Conservation of Habitats and Species Regulations 2017* and receive legal protection under Part 3, regulation 41, paragraph (1), which states:

> A person who —

> (a) deliberately captures, injures or kills any wild animal of a European protected species,
> (b) deliberately disturbs wild animals of any such species,
> (c) deliberately takes or destroys the eggs of such an animal, or
> (d) damages or destroys a breeding site or resting place of such an animal,

> is guilty of an offence.

Note: The offence of damaging or destroying a breeding site or resting place does not include the word 'deliberately', but is an *absolute* offence that does not require any fault elements to be proved to establish guilt.

Part 3, regulation 41, paragraph (2) states that disturbance of animals includes any disturbance which is likely:

(a) to impair their ability —
    (i)   to survive, to breed or reproduce, or to rear or nurture their young, or
    (ii) in the case of animals of a hibernating or migratory species, to hibernate or migrate; or
(b) to affect significantly the local distribution or abundance of the species to which they belong.

The legislation protecting all bat species discourages a good deal of amateur naturalists from looking for bat roosts, because inspection results in disturbance which is legally prohibited.

At the time of writing, the legislation that is most prohibitive to amateur naturalists is the *Wildlife & Countryside Act 1981*, which states:

Subject to the provisions of this Part, a person is guilty of an offence if intentionally or recklessly —

(b) he disturbs any such animal while it is occupying a structure or place which it uses for shelter or protection.

When this legislation came into force, the disturbance offence meant that in order to survey for bats, and indeed any of the species that were now protected, some sort of 'get out of jail free card' had to be created by the Department for Environment, Food & Rural Affairs (DEFRA), and so the 'Science & Education Licence' was born.

This Licence allowed the holder to survey for bats for the purposes of Science and Education on the condition that data collected was released to the appropriate recording scheme (in the case of bats, the Local Environmental Records Centre (LERC)).

The new legislation meant that only a limited group of people could now search for any of the wide range of species that received legal protection, a list spanning every faunal group from invertebrates, through amphibians and reptiles, fish, birds and mammals.

Amateur naturalists might in theory be breaking the law if they encountered a legally protected species when they were looking for something else entirely. Effectively they should know the ecology of every species in detail, simply so they could plan to avoid them!?

And so being a naturalist was suddenly made much more exciting by the prospect of being arrested!

This all came to a head when a new method for establishing dormouse *Muscardinus avellanarius* presence using nest-tubes was published by Chanin & Woods (2003). Nest-tubes were cheap, lightweight, easy to install and simple to check, but dormice had the same level of legal protection afforded to bats, and very few people had a dormouse survey licence, so the uptake was relatively low.

To counter this effect, English Nature (now Natural England) decided to take a pragmatic approach, and in the second edition of the *Dormouse Conservation Handbook* (Bright *et al.* 2006) they included this paragraph:

Inspecting nest boxes (and nest tubes) requires a licence from English Nature or the Countryside Council for Wales in areas where dormice are already known to be present, BUT if boxes or tubes are put out speculatively to detect presence, this in itself does not require a licence, but a licence is essential once the first dormouse has been found.

In summary, what this means is that if you know dormice are present (and you can check in advance with your LERC who hold all the records submitted by licence-holders), you should not be messing them about without a licence, but if you do not know whether they are present, and nor does anyone else, then you can have a look to see if they are. This allows anyone to go out and look for dormice whether they have a licence or not, on the understanding that as soon as the species is encountered the unlicensed naturalist will withdraw, inform the LERC themselves, and then seek qualified licensed advice if they want to continue to study the population they have discovered.

The same principle has been applied to a wide range of other species, including bats (see BTHK 2018).

Not having a licence does not therefore mean that you should not look for new roosts, it simply means you should not interfere with a known roost when you know that bats are occupying it.

There is no doubt that some inspections using torches do disturb roosting bats. Inserting an endoscope into a PRF is intrusive and some bats do react aggressively to the intrusion. When groups of females are present with young their response is at least to become agitated and they may exhibit significant panic. Notwithstanding: **a)** bats do return to roosts year after year despite intrusive surveillance (including endoscopy); **b)** bats do occupy artificial roost boxes despite being repeatedly caught and handled; **c)** bats do occupy cliffs despite them being popular with climbers; and, **d)** the presence of cavers (US – spelunkers) does not cause the wholesale abandonment of caves and mines by bats.

While it is accepted that Wikipedia is probably not the best resource for legal advice, it does summarise neatly that in criminal law recklessness may be defined as the state of mind where a person deliberately and unjustifiably pursues a course of action while consciously disregarding any risks arising from such action.

The reason this book was begun was that our knowledge has gaps big enough to drive a bus through, and right now no one really knows what they are looking for; that is justification for setting out to look and record what is found in order to share the data for the good of the bats. If bats are then encountered, disturbance should be confined to the minimum required to achieve a biological record,[3] and that biological record is the justification for the disturbance.

From a project perspective, a licensed inspection is no less disturbing to the bat than an unlicensed one – the bats do not ask to see the licence and then relax when it is produced. If all you are doing is randomly endoscoping with no intention of recording then that might reasonably be considered reckless and unjustifiable, regardless of whether you have a licence or not.

> **Note:** At the time of writing the UK has recently left the European Union. This will have an effect upon the legislation that protects bats: it may just be a matter of 'rebranding' or it may be much more dramatic. This could not be anticipated and it is recommended that readers are aware that the legislation may change.

1.4 Approach

1.4.1 Evidence base

Every part of this book has an evidence base and the evidence is cited every time.

1.4.2 Scoping-in

This is the start of the project. The approach taken is therefore to 'scope-in' where there is certainty and to leave doubt where there is not.

1.4.3 Confidence in the evidence

The use of the words 'record' and 'report' will be used to denote confidence.

Recorded means that the BRHK project holds, or has had sight of conclusive evidence of a bat (identified or not) roosting in a specific situation.

Reported means that someone, somewhere (named or not) has stated in a book or other published media that they have encountered a bat (identified or not) roosting in a specific situation, but the BRHK project has not had sight of evidence that would corroborate this account.

In simple terms, recorded means that we can prove the account is accurate and true and reported means that while we hope the account is wholly accurate, it may be inaccurate if not wholly fictitious.

The review of the rocks is therefore limited to only those in which there is a record or report of bats roosting that is presented later in one of the three habitats.

1.4.4 Approach to the environment

The environment is considered at three scales, as follows: **1) macro-environment** – the rock the landform is composed of and the location of the landform, the latter of which determines the climate it is exposed to; **2) meso-environment** – the landform and the external characteristics of the PRF it holds; **3) micro-environment** – the internal characteristics of the PRF.

1.4.5 Approach to the bats

The macro-, meso- and micro-environments in which individual bat species have been recorded and reported roosting are identified. In addition, some early work that has significant value but may have historically (and entirely unintentionally) been misinterpreted or misapplied, is discussed with additional analysis as appropriate to explain specific aspects.

1.4.6 Surrogate species

Finally, the *Bat Tree Habitat Key* project (BTHK 2018) has proven that some common invertebrate species occupy the same niche position on environmental 'shelter' gradients as roosting bats. These species may then be used as surrogates to infer the suitability of the environment to a specific bat species during a simple visual inspection of a PRF. These limited data are summarised.

1.5 Format

Continuing the approach taken in the *Bat Tree Habitat Key*, each of the three habitats is presented in identical format and with identical content. This ensures that nothing is overlooked and all are given equal weight. The advantage of this approach is that it identifies: **a)** what has been confidently proven and thoroughly described; **b)** what is uncertain and needs further 'truthing'; and, **c)** what is not known at all.

In order to standardise the approach taken across the very different physical structures of the three habitats, the project recording forms are used as the format framework. The rationale is that these give a common structure with some parts being immediately familiar across all three habitats, and those that require more thought being visibly grouped to programme the semantic memory,[4] and to prepare it for episodic individual and personal reinforcement.

By structuring the chapters in line with the format of the recording forms, all the subsections will be of practical value and follow the order in which a PRF would be recorded in the field. Copies of the recording forms are provided in Tables 1.1–1.5.

Table 1.1 The *BRHK Rock Face Database* recording form.

BASIC RECORD	**WHO**	1st Recorder:		2nd Recorder:
	WHERE	Site name:		
		Grid reference:		
	WHEN	Date:		
	WHAT	Rock type (e.g. granite):		
		Is the land form Landform in: an escarpment / a crag / a tor / a gorge / a boulder or block / a cutting / a quarry		
		Habitat on/over the landform (phase 1):		
		Rock face height (m):		
		Rock face aspect (e.g. North):		
		Is the rock face angle: reclined / sheer / overhung		
		What is the bat occupying: a crack / a break / a flake / a mono / a bidoigt / a pocket / a bucket / an off-width		
		PRF height (measured from base of face) (m):		
		PRF entrance aspect (e.g. North):		
		PRF entrance angle (e.g. 90° use protractor):		
		PRF entrance height (m / cm):		
		PRF entrance width (m / cm):		
		ROOST ACCOUNT	Bat species:	
			Number of bats:	
			Roost position in relation to the entrance: above / in front / to the side / below	
			Distance from entrance (cm):	
			Awake or torpid:	
			Droppings:	
			Odour cues (i.e. smell): none / pleasant / not unpleasant / repellent	
ADVANCED RECORD	**WHAT**	**PRF INTERNAL**	Height (i.e. top of entrance to top of interior) (cm):	
			Width (i.e. inside entrance lip to the back wall in front) (cm):	
			Depth (i.e. bottom of entrance to bottom of interior) (cm):	
			Apex shape: a spire / a peak or wedge / flat / unknown	
		INTERNAL CONDITIONS Circle as many of the fields as are applicable	**Internal substrate:** glossy smooth / flowing bumpy / sandpaper rough / jagged	
			Internal cleanliness:	*Primary* – clean / waxy / stained / polished
				Secondary – messy / dusty with loose debris / muddy or silty / sludgy
			Internal humidity: dry / damp / wet / green algae	
			Competitors:	
	COMPREHENSIVE INSPECTION POSSIBLE (yes or no):			

Table 1.2 The *BRHK Loose Rock Database* recording form.

WHO	1st Recorder:				2nd Recorder:	
WHERE	Site name:					
	Grid reference:					
WHEN	Date:					

WHAT							
	Rock type (e.g. granite):						
	Landform type: scree / talus / blockfield						
	Landform composition – Proportion of surface occupied by:	Pebbles:	Chips:	Cobbles:	Plates:	Boulders:	Blocks:
	Landform origin: natural / artificial						
	Landform shape: cone / fan / other						
	Landform topography: flat / stepped / convex / concave						
	Landform aspect (e.g. North or N/A if it is simply a pile of rock):						
	Angle of slope:						
	Contiguousness: continuous / discrete within network / discrete and isolated						
	Habitat around and on the landform (phase 1):						
	% island vegetation present (i.e. is anything growing within the landform):						
	% moss cover:						
	% of clasts* without moss cover that hold lichens:						
	Entry points:	Obvious gaps into which a bat might climb			No obvious spaces; bat would have to burrow into substrate		
	ROOST ACCOUNT	Bat species:					
		Minimum number of bats:					
		Maximum distance from edge of landform to roost position (m):					
		Awake or torpid:					
		Roost depth (cm):**					
		Describe any visual field-signs that might have given the presence of a roost away just by looking at the landform:					

* A clast is a particle of rock derived by weathering and erosion (Kearey 1996).
** Only if being measured as part of specific academic study or justified by circumstances.

Note: There is no 'advanced record' section because at the present time there is no practical way of collecting the data without risking damage to the PRF and indeed the bat.

Table 1.3 The *BRHK Subterranean Rock Database* recording form – macro values.

WHO	1st Recorder:	2nd Recorder:
WHERE	Site name:	
	Grid reference of the primary entrance:*	
WHEN	Date:	
WHAT 1	**Macro – Rock type (e.g. granite):**	
	Macro – Landform: solution cave / sea cave / mine / railway tunnel / other:	
	Macro – Entrance topography: Does the landform open in from: a sheer face / sloping ground / level ground / a dell or pit / a cutting (tick as many as apply)	
	Macro – Habitat over the opening through which the surveyors entered (phase 1):	
	Macro – Number of 'bat-friendly' entrances (include shafts and chimneys/avens, etc.):	
	Macro – Does the landform hold a: free surface stream / sump stream / sump / lake / no water	

* The primary entrance is the one that offers the easiest access and the one that cave rescue would be most likely to use if they had to get you out, which is EXACTLY why you are recording it: in case you or anyone else gets into trouble!

> **Note:** The macro values will not be recorded again. Subterranean rock landforms do not change in the way that faces and loose rock might, and unlike tree roosts (which are created and destroyed by storms etc.) subterranean features are functionally permanent. To make the point, you might record a roost in a subterranean landform today that your great, great, great-grandchildren will go back and check in over 100 years' time; that is less likely to be true of a rock face and even less still with a loose rock landform.

Table 1.4 The *BRHK Subterranean Rock Database* recording form – meso and micro values for exposed surface and recessed roost positions.

WHO	1st Recorder:	2nd Recorder:	
WHERE	Site name:		
WHEN	Date:		

BASIC RECORD	WHAT 2.1	**Meso 1a – Is the roost in the:** threshold / dark-zone	**Lux value:**	
		Meso 1b – What is the Distance between bat/PRF and nearest entrance (m):		
		Meso 2 – Is the roost position/PRF on the: ceiling / left wall / right wall / face / floor		
		Meso 3a – Is the roost feature: on a route / at a node or junction / in a destination *Now go to Meso 3b …*		
		Meso 3b – Is the roost feature on a route in a: portal / aisle / passage / keyhole passage / adit / tunnel / crawl / squeeze *– if yes >> if no go to 3c*	**What is the ceiling height (m):** **What is the width between the walls (m):** **Is the roost feature:** In a wider section of a route (i.e. you can walk past it) / at the end of a cul-de-sac / in a face *Now go to Meso 4…*	
		Meso 3c – Is the roost feature in an: aven / shaft *– if yes >> if no go to 3d*	What is the diameter of the tube (m): *Now go to Meso 4…*	
		Meso 3d – Is the roost feature in a: hall / chamber / cave / cavern *– if yes >> if no go to Meso 4*	**What is the ceiling height (m):** **Width on the north/south axis (m):** **Width on the east/west axis (m):** *Now go to Meso 4…*	
		Meso 4 – On the map you are using to navigate, does the meso-environment feature have a name? *– if yes >> if no go to Micro 1*	**Write the name here:** *Mark the roost location on the map*	
	WHAT 2.2	**Micro 1 – Is the roost position in an exposed plane surface situation?** yes / no *– if yes complete the record below if no go to Micro 2a*		
		Bat hanging from: rock: flat surface, step or rib / brick: flat surface, step or rib / wood: crossbar – prop – cribbing / metal: beam, bearing plate or nail		
		Bat species:		
		Number of bats:	**Is the bat:** awake / torpid	
		Height of the roost position above the floor (m):	**Droppings:** yes / no *Now go to 'associates'…*	
		Micro 2a – Is the roost position or PRF in a simple recessed feature, such as a: pitch / cupola / manhole / vestibule / shelf / tent / ceiling pocket / wall pocket *– If the roost is in a double-recessed feature (e.g. a ceiling pocket in a cupola, or a wall pocket in a vestibule) the dimensions of the larger component are recorded below and those of the smaller component are recorded at 2b. If there is only an individual recess you may ignore 2b and go to 'associates'*		
		Bat species:		
		Number of bats:	**Is the bat:** awake / torpid	
		Height of the roost position above the floor (m):	**PRF entrance aspect:** down-facing / outfacing / up-facing from floor	
		Entrance dimension on longest axis (cm):	**Entrance dimension on shortest axis (cm):**	
		Internal front-to-back width (cm):*	**Droppings below the bat:** yes / no *Now go to 'associates'…*	
		Micro 2b – Is the roost position or PRF in a double-recessed feature such as a: ceiling pocket / wall pocket		
		Entrance dimension on longest axis (cm):	**Entrance dimension on shortest axis (cm):**	
		Internal front-to-back width (cm):*		
		Micro 3 – List any associate invertebrate species present in the immediate vicinity:		

* Imagine you are looking at the dimensions of wardrobes at ikea.com – this is the depth measurement that would give the third dimension.

Table 1.5 The *BRHK Subterranean Rock Database* recording form – meso and micro values for enveloping roost positions.

WHO	1st Recorder:	2nd Recorder:
WHERE	Site name: see Chapter 1	
WHEN	Date: see Chapter 1	

Meso 1a – Is the roost in the: threshold / dark-zone		**Lux value:**
Meso 1b – What is the Distance between PRF and nearest entrance (m):		
Meso 2 – Is the roost in the: ceiling / left wall / right wall / face / floor		
Meso 3a – Is the roost feature: on a route / at a node or junction / in a destination *Now go to Meso 3b …*		
Meso 3b – Is the roost feature on a route in a: portal / aisle / passage / keyhole passage / adit / tunnel / crawl / squeeze *– if yes >> if no go to 3c*	**What is the ceiling height (m):** **What is the width between the walls (m):** **Is the roost feature:** in a wider section of a route (i.e. you can walk past it) / at the end of a cul-de-sac / in a face *Now go to Meso 4…*	
Meso 3c – Is the roost feature in an: aven / shaft *– if yes >> if no go to 3d*	What is the diameter of the tube (m): *Now go to Meso 4…*	
Meso 3d – Is the roost feature in a: hall / chamber / cave / cavern *– if yes >> if no go to Meso 4*	**What is the ceiling height (m):** **Width on the north/south axis (m):** **Width on the east/west axis (m):** *Now go to Meso 4…*	
Meso 4 – On the map you are using to navigate, does the meso-environment feature have a name? *– if yes >> if no go to Micro 1*	**Write the name here:** *Mark the roost location on the map*	
Micro 1 – Is the roost feature sheltered within a simple recessed feature? yes / no *– If yes go to Micro 2, if no go to Micro 3…*		
Micro 2 – Circle the relevant recess type: pitch / cupola / manhole / vestibule / shelf / tent / ceiling pocket / wall pocket		
Recess entrance aspect: down-facing / outfacing / up-facing from floor	**Recess entrance dimension on longest axis (cm):**	
Recess entrance dimension on shortest axis (cm):	**Recess front-to-back width (cm):***	
Micro 3 – Which of the following is the bat occupying: a crevice / crack / break / fissure / pleat / alicorn / borehole / drain-hole / breakdown / choke / fill / other (please describe): *Now complete the record below …*		
PRF height above or below floor (m):	**PRF in:** rock / brick / wood / metal	
PRF entrance aspect: down-facing / up-facing / outfacing from a wall		
What is the entrance dimension on the longest axis (cm):		
What is the entrance dimension on the shortest axis (cm):		
Bat species:	**Number of bats:**	
Is the bat: awake / torpid	**Are droppings present in the PRF:** yes / no	
Where is the bat in relation to the roost entrance: above / in front / to the side / below *If you are licensed to use an endoscope go to Micro 4…*		
What is the distance between the entrance and the bat (cm):		
What is the internal height (i.e. top of entrance to top of interior) **(cm):**		
What is the internal width (i.e. inside entrance lip to the back wall in front*) **(cm):**		
What is the internal depth (i.e. bottom of entrance to bottom of interior) **(cm):**		
Is a comprehensive inspection possible: yes / no		
Internal humidity: aridly dry / surface darkened by damp / obviously wet with runs or droplets		
List any associate invertebrate species present inside:		

Left vertical labels: BASIC RECORD (WHAT 2.1, WHAT 2.3); ADVANCED RECORD (WHAT 2.4).

* Imagine you are looking at the dimensions of wardrobes at ikea.com – this is the depth measurement that would give the third dimension.

If the forms are compared, it will be noted that they encompass all four basic pieces of information required for a biological record, comprising: **1)** who; **2)** where; **3)** when; and, **4)** what.

Regardless of the form used, the 'who', 'where' and 'when' are all recorded in the same way. Only when the recorder gets to the 'what' does the task become more complicated.

1.5.1 Who

The 1st Recorder is the person doing the inspection. Ideally, they will be assisted by a 2nd Recorder who will be filling-in the form. In fact, the roles are unfair inasmuch as it is the 2nd Recorder's responsibility to ensure the recording form is accurate and complete.

This is a vital consideration: ALL the boxes should have a value in them, either an attribute circled or described, or a numerical figure. Blank-spaces leave room for doubt and suggest the record is incomplete.

Anyone can do an inspection; it takes steadfast discipline to complete the form.

1.5.2 Where

The location includes two values, comprising: **1)** the site name; and, **2)** the Ordnance Survey (OS) Grid Reference.

The site name should be taken from OS 1:25,000 sheets; however, the really important part is the OS reference because individual bat species will occupy specific niche positions on regulatory gradients.

These gradients will include: **1)** geology; **2)** average monthly range of air temperature; **3)** average monthly wind speed in knots; **4)** number of days per month with measurable rainfall; **5)** number of days per month with rainfall over 13 mm (½"); **6)** average monthly precipitation (mm).

With a 12-figure alphanumeric grid reference, we can consult online geological and climate databases that span decades, and build a map of where a specific species CAN occur, and where they are most likely to occur, in different periods.

Finally, in order to share what we find, we can convert the OS reference into latitude and longitude to allow comparison with wider European populations in similar climatic zones. This means that the data have a value outside the UK.

> **Note:** Some people prefer to be vague about their records in order to stop other less ethical and mannerly individuals from going and disturbing the roost. That is both fair and reasonable, provided the confidential record is submitted to the LERC. The BRHK project does not release the OS reference to anyone other than an academic institution, and ideally the record will be recorded to the full 12-figure alphanumeric code to enable review and detailed analysis.
>
> Regardless, if you put an endoscope into any PRF and do not record it, you are an irresponsible vandal and should be tarred, feathered and run out of town on a rail. It cannot be overstated: every time you put your endoscope into a PRF, at best all you are doing is disturbing bats (and anything else in the PRF), at worst you may be damaging the structure.
>
> If you are now attempting to justify why you have not been diligently recording in such a way as to assist in scientific progress, the test is to attempt to explain what you have been doing as though you were speaking to an audience of bats: how will it benefit them?

1.5.3 When

The date the PRF was recorded.

1.5.4 What

The 'what' is all the meso- and micro-scale environmental values that combine to make the environment suitable or unsuitable for occupation as a roost by a specific bat species.

These are the subtleties that deliver the niche positions exploited by the individual bat species on the wider environmental gradients. It is this knowledge that allows us to play the game.

Any idiot can physically search for a bat and find it in a PRF: all that takes is the right tools and persistence. The trick is to predict which bat species will be present, when and where before you even get out the endoscope.

Do that and you become a player.

Players get the full record, enter it into Excel and build a database they can use to assess the suitability of each new situation, and by so doing they learn where the niche position is on the overall tolerance range (i.e. where the bat is most comfortable within the range of what it can physically endure). Players practise constantly; they return to the same PRF on a regular basis and record the internal environment regardless of whether bats are present or not; they are interested in the bats and not their own self-aggrandisement.

Players compete against the habitat, against the bats and against themselves.

Players approach a survey as they would a shot in a game of pool.

They consider the habitat as though it were the table.

They name the ball; ironically, in this case the ball is in fact the bat species that they know exploits that habitat, and that they also know is on the table because they know it occurs in that county.

They then name the pocket: the specific PRF that the specific landform offers.

Their knowledge of the micro-environment focuses their aim.

And every shot is exciting, a test of their knowledge …

Note: This is not a 'coffee-table' book. Even if that was what we were aiming for, two COVID lockdowns have frustrated the production of the book at every turn, and although we know where we could get some stunning photos, we simply have not been able to get out to take them.

However, in the end the amount of information that had to go into each chapter meant that space for photographs was limited. There are therefore a lot more photographs of Potential Roost Features without bats in than there are of PRF with bats in. This is because when the features are stuffed full of bats you cannot see the feature properly.

We also specifically did not want this to be a competition – if you have a landform with only one bat species in it, or only one bat, it is still an interesting landform and you should still map it.

The next point is that not all the features are the best examples we could show you, the 'textbook' obvious examples. What we have tried to do is to give more subtle examples because the obvious ones are, well, obvious, and this is supposed to be a helpful field guide.

Notwithstanding, the rock faces are biased to Wales and Yorkshire, and everything else is biased to the south-west of England or wider Europe. In addition, the classification frameworks are not perfect and there is no doubt that you absolutely ARE going to find bats in features that are not in this book.

This book is therefore going to need updating in the future.

A journey of 1,000 miles starts with a single step. But to take that step you first have to decide on a direction. This is a stab at a direction. It is not the destination. But one day *Homo sapiens* will reach the destination and look back on the journey, wiser for the experience.

CHAPTER 2

An introduction to rock

Henry Andrews & Steven Hopkins

In this chapter	
Preamble	The rationale for considering rocks before bats; a confession and the primary sources of evidence and information; the chapter layout and content
Separating rock from stone	The distinction between rock and stone
The 23 rocks for which there are records and reports of bats exploiting plus five more to explore	A list of the 23 rocks, of which there are associated records or reports of roosting bats in 20, and the other three may be worth investigating
The broad geographical distribution of the rocks	A broad and very general summary of the broad distribution of the 23 rocks across the British Isles
The origin and grain size of the rocks	How the rocks were formed and the size of the grain
The rock landforms	An introduction to: **1)** escarpments; **2)** crags; **3)** tors; **4)** gorges; **5)** boulders and blocks; **6)** cuttings; **7)** quarries; **8)** screes; **9)** talus; **10)** blockfields; **11)** solution caves; **12)** sea caves; **13)** mines; and, **14)** railway tunnels
Which of the rocks form specific natural landforms in the British Isles	A summary of the natural landforms offered by each of the rocks in the British Isles
Which of the rocks exist as specific artificial landforms in the British Isles	A summary of the rocks that have been worked to leave behind specific artificial landforms in the British Isles

2.1 Preamble

A good starting point in understanding how bats use rock is to understand what rock can offer.

The first of the four BRHK objectives is:

The provision of a handbook that describes all the broad descriptions of all the features commonly encountered in and on natural and semi-natural surface, subsurface and subterranean rock habitats. The purpose of the handbook will be to inform targeted searches for Potential bat Roost Features (PRF).

This chapter will start fulfilling that objective by identifying the rocks that we already know have some value to bats, and others that may be worth investigating. As the project progresses, we will be looking at such things as: **a)** whether different rocks behave differently in different situations; **b)** whether any specific rock forms a particular PRF better than all the others; **c)** whether different rock types offer simple or more complex PRF; and, **d)** whether different rocks have different properties and therefore whether they offer structurally the same PRF with different internal environments.

The idea is to see what the rocks can actually deliver in terms of PRF, because it is the rocks and the PRF that a surveyor can see in the day and they are therefore the basis that any prediction of roost presence has to work up from.

It is logical to hypothesise that as each individual bat species has evolved to occupy different niche positions in the overall gradients delivered by rock, for more than one species to exploit a particular rock it would have to offer a relatively wide range of roost environments. It is also logical to hypothesise that there might be individual rocks that offer an individual environment that one bat species is better adapted to exploit than any other. Finally, it is logical to hypothesise that there will be some environments offered that are less attractive to all bat species.

However, working up an initial set of hypotheses into a theory is complicated by the fact that different geological conditions often occur in different geographical situations, and this means that different sorts of rock are offered to different combinations of bat species.

The geographic location of the rocks determines the macro-environment and we can reasonably suppose that granite in the Scottish Highlands will be subject to a significantly different climate from granite in Cornwall. As the geographic distribution and structure of these rocks is different, so is the meso-environment they offer in terms of landforms and PRF, and this (coupled with factors such as aspect[5] and porosity)[6] influences the micro-environment inside the PRF.

This chapter is an introduction to the rock types that have sufficient structural strength to form long-lasting PRF.

The chapter starts by separating rock from stone. After that, a list of all the rock types for which there is already a recorded or reported association with roosting bats identified. Thereafter, summary descriptions are given in respect of: **a)** the broad geographical distribution of the rocks; **b)** the origin and grain size of the rocks; **c)** whether the rock forms natural landforms and what their broad characteristics are; and, **d)** whether the rock has now or has ever been worked to offer artificial landforms and what their broad characteristics are.

2.1.1 Confession

The primary author of this book is not a geologist, does not rock climb and his caving experience is limited to systems used for school outings. To give perspective to the climbing community, each new page of *Hard Rock* (Parnell 2020) was met with a wide-eyed head-shake and a reaction ranging from a murmur of 'these people are unhinged …' through an immediate page-turn like a child finding a large photograph of a spider. To give the same context to the caving community, a copy of *Blind Descent* (Tabor 2011) provided by our publisher was met with a similar reaction, only with more '… whaaaaaaaat?!?!?'

In order to pitch the chapter at a level that will introduce rocks to amateur naturalists and professional ecologists, without irritating climbers and spelunkers, the narrative leans heavily on five published texts, comprising:

» Garlick, S. 2009. *Flakes, Jugs, and Splitters: A Rock Climber's Guide to Geology (How To Climb Series)*. FalconGuides, Richmond.

» Kearey, P. 1996. *The New Penguin Dictionary of Geology*. Penguin, London.

» Parnell, I. 2020. *Hard Rock: Great British Rock Climbs from VS to E4*. Vertebrate, Sheffield.

» Waltham, A., Simms, M., Farrant, A. & Goldie, H. 1997. *Karst and Caves of Great Britain: Geological Conservation Review Series No. 12*. Chapman & Hall, London.

» Wilson, K. 2007. *Classic Rock: Great British Rock Climbs*. Bâton Wicks, Sheffield.

The following online resources were also invaluable, comprising:

» ukclimbing.com – The most outstanding resource of anything anywhere and I wish we had a bat website like it

» bgs.ac.uk – The website of British Geological Survey, a mine of information (see what we did there)

» wikipedia.com – What modern resource would be complete without a visit to Wikipedia; please visit now and make a donation

My co-authors and those of you who also climb or spelunk will be the heroes of this story.

2.2 Separating rock from stone

To begin at the beginning, *The Penguin Dictionary of Geology* (Whitten & Brooks 1979) is clear that mineral matter that forms part of Earth's crust is rock. The word 'stone' is used to describe extracted material such as limestone, building stone, road stone, etc., and should not be used as a synonym for rock. Furthermore, where a pebble has a natural origin, small stones do not.

While it might appear reasonable (if pedantic) to describe a natural crag, tor, sea-cliff, scree and a phreatic cave passage through limestone as rock, and artificial cuttings, quarry faces, mines and tunnels as stone, in fact, in an artificial situation (unless the face has been dressed with bricks and mortar, etc.) what has been exposed is rock.

Therefore, escarpments, crags, tors, gorges, boulders, cuttings, quarries, screes, talus, blockfields, solution caves, sea caves, mines and railway tunnels are all rock features. Only quarry and mine *spoil* that replicates screes, talus and blockfields and the brick lining of a railway tunnel would be stone.

2.3 The 20 rocks for which there are records and reports of bats exploiting plus three more to explore

So far there appear to be 20 players in the bat rock hall of fame, comprising: **1)** basalt; **2)** chalk; **3)** mudrock; **4)** dolerite; **5)** diorite; **6)** gneiss; **7)** granite; **8)** greywacke; **9)** gritstone; **10)** limestone; **11)** mudstone; **12)** quartzite; **13)** rhyolite; **14)** sandstone; **15)** schist (including variants such as mica-schist); **16)** shale; **17)** siltstone; **18)** slate; **19)** tonalite; and, **20)** tuff.

There are also at least three more that will be worth investigation, comprising: **21)** andesite; **22)** coal; and, **23)** gabbro.

2.4 The broad geographical distribution of the rocks

Different rocks occur in different locations in the British Isles. From the perspective of climbers, cavers, quarrying and mining, as a very general order, we see andesite, gabbro, gneiss, granite, rhyolite, sandstone and schist in the Scottish Highlands, passing across basalt and greywacke before giving way to gneiss and an abundance of rhyolite in the north of England and Wales (with mudstone/siltstone[7] on the Yorkshire coast and quartzite on the sea escarpments of Wales), down into the diorite, limestone, mudstone/siltstone, sandstone and gritstone that continues through the English Midlands, which in turn gives way to the limestone of south-west England and Wales, chalk (interspersed

Table 2.1 A summary of the origin and grain size of the 20 rocks that appear to be important to roosting bats plus three more that may be worth investigating (the latter identified by being shaded grey).

ROCK	ORIGIN	GRAIN	EVIDENCE OF VALUE TO ROOSTING BATS
Andesite	Igneous[1]	Fine	—
Basalt	Igneous	Fine	1 SSSI notification for a mine in Wales*
Chalk	Biogenic[2]	Fine	Stebbings (1988); 13 SSSI notifications for mines and tunnels in England*
Mudrock	Sedimentary[3]	Fine	Ancillotto et al. (2014) – N.B. study performed in Italy
Coal	Sedimentary[4]	Fine	—
Dolerite	Igneous	Medium	Shewring (2015)
Diorite	Igneous	Coarse	1 SSSI notification for a tunnel in England*
Gabbro	Igneous	Coarse	—
Gneiss	Metamorphic[5]	Medium to coarse	Michaelsen et al. (2013); Shewring (2015)
Granite	Igneous	Coarse	Barrett-Hamilton (1910 – see Myotis daubentonii account in Glen Dochart); 1 SSSI notification for a mine in England;* Shewring (2015)
Greywacke	Sedimentary[3]	Fine to coarse	Shewring (2015)
Gritstone	Sedimentary[3]	Fine to coarse	BRHK Rock Face Database; Shewring (2015)
Limestone	Biogenic[4]	Fine to coarse	Ransome (1968); Ransome (1990); Billington (2004); S. Dyer, pers. comm., June 2013; BRHK Rock Face Database; BRHK Subterranean Database; 76 SSSI notifications for caves, mines and tunnels in England and Wales*; Shewring (2015)
Mudstone	Sedimentary[3]	Fine	7 SSSI notifications for mines in England and Wales*
Quartzite	Metamorphic	Medium	1 SSSI notification for a mine in Wales*
Rhyolite	Igneous	Fine	Shewring (2015)
Sandstone	Sedimentary[3]	Fine	Billington (2000); BRHK Rock Face Database; 13 SSSI notifications for caves, mines and tunnels in England and Wales*
Schist	Metamorphic	Fine to medium	BRHK Rock Face Database; Shewring (2015)
Shale	Sedimentary[3]	Fine	1 SSSI notification for a mine in Wales*
Siltstone	Sedimentary[3]	Fine	2 SSSI notifications for mines in Wales
Slate	Metamorphic	Fine	Mitchell (2008); Parker et al. (2013); 3 SSSI notifications for mines in England and Wales*
Tonalite	Igneous	Course	1 SSSI notification for a tunnel in England*
Tuff	Igneous	Fine	BRHK Rock Face Database; 2 SSSI notifications for mines in Wales*

[1] Igneous rock is solidified from molten or partially molten material and can be extrusive and fine-grained (e.g. andesite, basalt, rhyolite and tuff) that has erupted, or intrusive and coarse-grained (e.g. diorite, gabbro, granite and tonalite) formed within Earth's crust (Garlick 2009).

[2] Solidified from tiny particles of marine organisms.

[3] Solidified from tiny particles that have settled on a sea bed.

[4] Formed from the accumulation of organic material in tropical prehistoric swamps and rainforest.

[5] Metamorphosed from one rock to another through the application of significant heat or pressure, e.g. sedimentary mudstones being metamorphosed into slate.

* See Appendix A.

with nodules of chert) in the south and south-east of England, china/ball clay in Devon and Cornwall, and finishing with more granite, sandstone and schist.

Northern Ireland begins with basalt and tuff. Moving south we see gneiss, schist and quartzite giving way to an abundance of limestone that continues south across the greater proportion of the Republic of Ireland before finally giving way to sandstone. Gabbro spans the border on the east coast, and granite also occurs scattered in the east and west.

Coal is considered separately because it is usually limited to subsurface working. Naturally, its distribution is reflected in the British Coalfields, ranging from the Clyde in Scotland, those in the north-east and north-west of England, Staffordshire, Bristol, South Wales and Kent.

2.5 The origin and grain size of the rocks

Keeping things simple, Table 2.1 gives a basic summary of the origin and grain size of each of the rocks and is constructed from Kearey (1996), Wikipedia and a general web search.[8] The evidence that the rock is used by bats is referenced in column 4.

2.6 Rock landforms

Typical rock landforms found in the British Isles are discussed below:

» **Escarpments** – Escarpments comprise a long and sheer cliff separating flat land at different levels. The plateau above and the land at the foot of the cliff are broadly level or have a gentle slope. Examples include Arthur's Seat in Holyrood Park, Edinburgh (see Photo 2.1), and Stanage Edge which spans Derbyshire and Yorkshire (see Photo 2.2).

> **Note:** In the context of this project, escarpments include cliffs that open onto the sea.

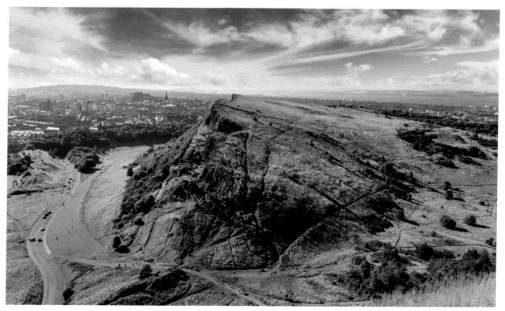

Photo 2.1 Arthur's Seat. (© S-F/Shutterstock)

Photo 2.2 Stanage Edge. (© Steve Cymro/Shutterstock)

» **Crags** – *The Penguin Dictionary of British Natural History* defines a crag as 'a steep irregularly shaped rock or outcrop of rock' (Fitter & Fitter 1967). Climbers apply the word crag to encompass any large exposure of free-standing and more-or-less bare stone that is suitable for rock climbing. However, the full and correct term is 'crag and tail', the form of which is sort of whale-shaped with rock promontory body leading into a tail of rock fragments and soil. An excellent example is Almscliffe Crag in Yorkshire (see Photo 2.3).

   Crags are the result of erosion by glaciers. As the glaciers passed over the ground, they removed the weaker materials leaving behind harder rock as a series of protrusions, which may span mountains through outcrops of a similar size to what we would think of in the context of tors. The mountain crags hold cliffs, but they also hold buttresses, gullies, caves and every other rock feature imaginable.

Photo 2.3 Almscliffe Crag. (© Jez Campbell/Shutterstock)

» **Tors** – These are effectively mini-crags, but in contrast to a crag which will be vegetated on at least one side and even on the top, a tor is entirely exposed on all sides and on the top. Tors are the result of erosion glacial movement and then weathering. Haytor and Hound Tor on Dartmoor, Devon are good examples (see Photos 2.4 and 2.5).

Photo 2.4 Haytor.

Photo 2.5 Hound Tor.

» **Gorges** – A gorge is a narrow valley between hills or mountains. They are the result of water erosion and typically have an ephemeral or perennial stream in the base. Gorge exposures are typically broken and interspersed with vegetated soil. Cheddar Gorge in Somerset is an obvious dry example (see Photo 2.6), as is the Valley of the Rocks in Devon, but the gorges of the East and West Lyn rivers are more typical of gorges the reader is likely to encounter (see Photo 2.7).

Photo 2.6 Cheddar Gorge. (© stocker1970/Shutterstock)

Photo 2.7 The River Lyn gorge.

» **Boulders and blocks** – A boulder is a rounded rock with a diameter of over 256 mm and a block is an angular rock with a diameter over 256 mm (Kearey 1996). Boulders and blocks may be the result of freeze–thaw weathering in mountainous situations, but in lowland situations they are typically found associated with a larger landform as the result of glacial erosion. Escarpments, crags and tors are formed by the movement of a glacier over an exposed rock outcrop. The glacier has advanced up smoothing one side into a ramp, and depositing the 'shaved-off' stone either at the base of the cliff on the other side (as in the one in Photo 2.8 which is below a tor) or further on (see explanation in Garlick 2009). Where a boulder is isolated in the landscape it may be a glacial erratic that has been transported by a glacier, and then left behind after the glacier melted.

» **Cuttings** – According to Wikipedia, 'the term cutting appears in the 19th century literature to designate rock cuts developed to moderate grades of railway lines, and are defined as: a passage cut for the roadway through an obstacle of rock or dirt'. Some cuttings are effectively artificial gorges wide enough to allow a railway line to pass through, as in that shown in Photo 2.9. Many in the UK are redundant and have been adopted as cycle-routes.

Photo 2.8 A granite boulder on Dartmoor.

Photo 2.9 A double-faced railway cutting. (© Frimu Films/Shutterstock)

» **Quarries** – A quarry is an open pit from which stone or ore is won for use in construction, manufacturing and even pharmaceuticals. Rock quarries are developed by controlled explosions to release the rock from the working face. In many situations the stability of the working face is maintained by working the face back in a reclined progression. This may result in a series of distinct benches or steps (see Photo 2.10) or a more-or-less smooth ramp (see Figure 2.1). In other situations, the face stone is stable enough for working in a sheer cliff (see Photos 2.11 and 2.12). In all cases, the result is effectively an artificial escarpment.

A single quarry may encompass artificial escarpments, tors, blocks and boulders. It is not uncommon for such a quarry to also encompass some or all of the following: cuttings, artificial talus, scree, natural caves, and mine entrances. Furthermore, in recent decades quarrying companies have sought to retain and even create sheltering features within worked-out quarries.

Photo 2.10 A benched quarry face.

Figure 2.1 Reclined smooth ramp-like faces.

Photo 2.11 A shallow limestone quarry with one continuous sheer quarry face.

Photo 2.12 A sheer quarry cliff.

» **Screes** – In their natural form, screes are typically triangular and sloping accumulation of loose and mobile fragments of rock; widest at the base and narrowing
up the slope in a fan or cone, and the result of weathering and erosion (Fitter &
Fitter 1967; Kearey 1996). Screes may be entirely bare (see Photo 2.13) or partially
vegetated (see Photo 2.14). In addition, artificial screes may also be encountered
in quarries, where their shape can be triangular (usually as a result of a landslip)
or one long piled slope (where they are the result of intentional landscaping) (see
Photo 2.15).

Photo 2.13 Natural bare scree. (© Matyas Rejak/Shutterstock)

Photo 2.14 A natural scree that is still mobile despite encroaching vegetation.

Photo 2.15 Artificial limestone scree in a Mendip quarry.

Photo 2.16 Granite talus on Dartmoor.

» **Talus** – Many texts consider 'talus' to be synonymous with 'scree' but in the context of this book it will be treated separately and used specifically to denote a landform that is dominated by blocks and boulders that range from washing-machine- to car-size. In addition to the size of the individual components, talus also differs from scree by a degree of structure and stability – inasmuch as some of the material may be interlocked and therefore static. In its natural form, talus is the result of catastrophic collapse and landslide. In the artificial form it is the result of quarrying and mining, as in the granite example in Photo 2.16.

» **Blockfields** – Blockfields comprise both boulders and blocks (and smaller rocks) spread across level or gently sloping ground. Although the size of the rocks is the same, in addition to fundamental differences in their formation (which are described in Chapter 4), blockfields are physically different from talus by their more-or-less level nature (a talus always has a sloping face). As with screes and talus, there are natural and artificial forms of blockfields. The granite example in Photo 2.17 is the result of historical quarrying and spreads over a wide area.

Photo 2.17 Artificial limestone blockfield in a Mendip quarry.

Photo 2.18 The entrance to a solution cave.

Photo 2.19 The entrance to a sea cave. (© Helen Hotson/Shutterstock)

» **Solution caves** – The term 'solution cave' will be used generally to mean any inland natural subterranean situation that was formed by the percolation of water through the rock, and which has resulted in a passage that extends sufficiently far into the rock to offer an entirely dark environment that an individual adult *Homo sapiens* could enter and take shelter from the weather outside. The example shown in Photo 2.18 is in Burrington Combe.

» **Sea caves** – Regardless of the mechanism of formation, the BRHK project considers all natural caves that open onto beaches as sea caves (i.e. caves that open onto the sea, but are distinct from marine caves in that sea caves are not tidally flooded to the ceiling). An example is provided in Photo 2.19.

» **Mines** – In the context of this book, the word 'mine' will be taken generally to mean any artificially formed situation that enters sufficiently far into the rock substrate to offer an entirely dark environment that an individual adult *Homo sapiens* could enter and take shelter from the weather outside in total darkness, even on days of full sun. Photo 2.20 illustrates a typical horizontal adit/passage entrance, and Photo 2.21 illustrates a typical shaft.

Photo 2.20 The entrance to an abandoned mine. (© Fredy Thuerig/Shutterstock)

Photo 2.21 A typical mineshaft.

» **Railway tunnels** – And so finally we come to railway tunnels which are a rock landform, although they may have a proportion of their face dressed with bricks, as illustrated in Photo 2.22.

In the late eighteenth century and throughout the nineteenth century the bulk of industrial transportation went by rail. This required level running gradients. In order to pass through hilly and mountainous areas such as South Wales and the Pennines, long tunnels had to be dug. Most of these tunnels are now redundant, but many still remain accessible, often adopted as cycle paths.

Railway tunnels have refuge manholes along their lengths which are staggered on both sides (i.e. they are not opposite each other but offset). These manholes are typically supported on their sides and roof with some degree of stonework, but with the back wall bare rock, as shown in Figure 2.2.

Photo 2.22 A railway tunnel – stone at the base and bricked over the arch.

Figure 2.2 Refuge manholes: stone and brick-lined on the sides and top but with the back wall bare rock.

2.7 Which of the rocks form specific natural landforms in the British Isles

A summary of the natural landforms offered by the rocks is provided in Table 2.2. The table was constructed using photographs in Wilson (2007), Parnell (2020), personal experiences and a great deal of time searching the Internet. Where there is a tick there is photographic evidence, where there is a question mark the searches failed to find any evidence.

> **Note:** Coal is specifically not included because we are confident that, without human intervention, in the British Isles it does not form any of the landforms with which bats are associated. In addition, the tabulation is specific to the British Isles: we know basalt does some wonderful ravines elsewhere in the world, but are there any examples in the UK?

Table 2.2 A summary of the natural landforms offered by the rocks.

| ROCK | ROCK FACES | | | | | LOOSE ROCK | SUBTERRANEAN | |
	Escarpment (including sea-cliffs)	Crag	Tor	Gorge	Boulders & blocks	Scree, talus & blockfield	Solution cave	Sea cave
Andesite	✓	✓✓	?	✓	?	✓✓	?	?
Basalt	✓	?	?	?	?	?	?	?
Chalk	✓✓	?	—	?	—	?	?	✓
Clay	?	?	?	?	?	—	✓	?
Dolerite	✓	✓✓	?	✓	?	✓	?	?
Diorite	?	?	?	?	?	?	?	?
Gabbro	✓	✓✓	?	✓	?	✓	?	?
Gneiss	✓✓	✓✓	?	✓	✓✓	✓✓	?	?
Granite	✓	✓✓	✓✓	✓	✓✓	✓	✓	✓
Greywacke	✓	✓✓	?	✓	?	?	?	?
Gritstone	?	✓✓	?	✓	?	?	?	?
Limestone	✓✓	✓✓	✓	✓✓	?	✓✓	✓✓	✓✓
Mudstone	✓	?	?	?	?	?	?	?
Quartzite	✓✓	✓	✓	✓	?	?	?	✓
Rhyolite	✓	✓✓	?	✓	?	✓✓	✓	?
Sandstone	✓	✓✓	?	✓	✓✓	✓✓	✓	✓
Schist	✓	✓✓	?	✓	?	?	?	?
Shale	?	?	?	?	?	?	?	?
Siltstone	?	?	?	?	?	?	?	?
Slate	✓	✓✓	?	✓	?	✓✓	?	✓
Tonalite	?	?	?	?	?	?	?	?
Tuff	?	✓	?	?	?	?	?	?

✓✓ – Examples of the landform occur in superabundance, both in number and individual scale.

✓ – Examples of the landform do exist but the landform does not appear to have a significant association with that rock.

? – No examples of the landform are known.

2.8 Which of the rocks exists as specific artificial landforms in the British Isles

A summary of the rocks that exist as artificial landforms is provided in Table 2.3. This is the result of personal experience, web searches and accounts of SSSI in England and Wales (see Appendix A).

Table 2.3 A summary of the artificial landforms that offered by the rocks.

ROCK	ROCK FACES		LOOSE ROCK	SUBTERRANEAN	
	Cutting	Quarry	Scree, talus & blockfield*	Mine	Railway tunnel
Andesite	?	✓	?	?	?
Basalt	?	✓	?	✓	?
Chalk	✓	✓	?	✓	?
Clay	?	✓	?	✓ – for the rock itself	?
Coal	?	✓	?	✓ – for the rock itself	?
Dolerite	?	✓	?	?	?
Diorite	?	✓	?	✓	✓
Gabbro	?	✓	?	?	?
Gneiss	?	?	?	?	?
Granite	?	✓	✓	✓ – for tin, copper, zinc, iron and lead	?
Greywacke	?	✓	?	?	?
Gritstone	?	✓	?	?	?
Limestone	✓	✓	✓✓	✓ – for the rock and also for ochre, copper and iron	✓
Mudstone	?	✓	?	✓	?
Quartzite	?	✓	?	✓	?
Rhyolite	?	✓	?	?	?
Sandstone	✓	✓	?	✓ – for iron	?
Schist	?	✓	?	?	?
Shale	?	✓	?	✓	?
Slate	?	✓	?	✓ – for the rock itself	?
Tonalite	?	?	?	✓	✓
Tuff	?	✓	?	✓	?

* In their artificial form screes, talus and blockfields are typically encountered within quarries.

Note: Cuttings and tunnels may pass through and rock but at the present time it is unknown exactly which have sufficient structural integrity to be found in a natural 'undressed' state with the bare rock exposed, and which require support by being shored up or dressed with stone.

Rock faces: characterising and recording the landforms and the Potential Roost Features they may hold

Henry Andrews, Robert Bell, James McGill & Hal Starkie

In this chapter	
Preamble	Recap of the previous chapter and relevant project objective; how Chapter 3 dovetails into Chapter 2 and how it contributes to achieving the project objectives; recap of the component parts of a biological record – who, where, when and what; and, introducing the component parts of 'what' in the context of rock face landforms
What	Introducing the meso-environment values; Introducing the rock face roost account – Bats, droppings and odour cues; Introducing the micro-environment values
What 1: recording the rock face meso-environment values	The rock type; The landform; The habitat on/over the rock landform; The rock face height; The rock face aspect; The rock face angle; The PRF type; The PRF height; The PRF entrance aspect; The PRF entrance angle; The PRF entrance height and width
What 2: recording the rock face roost account	The bats; droppings; odour cues
What 3: recording the rock face micro-environment values	The internal dimensions; The apex shape; The internal substrate; The internal humidity; The competitor species; Recording whether a comprehensive inspection was possible
The rock face bat species summaries	Barbastelle *Barbastella barbastellus* rock face roosting account; serotine *Eptesicus serotinus* rock face roosting account; Daubenton's bat *Myotis daubentonii* rock face roosting account; whiskered bat *Myotis mystacinus* rock face roosting account; Natterer's bat *Myotis nattereri* rock face roosting account; Common pipistrelle *Pipistrellus pipistrellus* rock face roosting account; Soprano pipistrelle *Pipistrellus pygmaeus* rock face roosting account; Brown long-eared bat *Plecotus auritus* rock face roosting account
The point	Finishing your career as a player and not a disillusioned hack

3.1 Preamble

In Chapter 2 the different rocks the reader might encounter in their part of the British Isles were identified. The landforms made of those rocks were then described so that they might be identified in the field.

The next step is to begin looking at Project Objective 2, which is:

The collation of published descriptions of bat roosting ecology on, among and in natural and semi-natural rock habitats.

We begin this process in Chapter 3 with rock-face habitats, and find that collating published accounts is a simple task because there is very little to read. All is not lost, however, because sufficient material is available to motivate us to look in the first place, and this material gives us some direction.

Applying the rationale set out in Chapter 1, the framework for the data review and the chapter layout will follow the order of the standardised *Bat Rock Habitat Key (BRHK) Rock Face Database* recording form. This will mean that the information read, collated and presented will have a practical application in the field.

In Chapter 1 we looked at the 'who', 'where' and 'when' of a biological record. Among this information, the 'where' allows us to identify the macro-environment in terms of the climate and geology.

This chapter covers 'what' else we need to record for a useful record of a specific rock face. Obviously, if an inspection of a Potential bat Roost Feature (PRF) encounters bats and/or droppings this immediately confirms it is a roost, but what if the PRF is empty: does that mean it is not suitable to hold roosting bats? Answering this question is achieved by a comparison of the environmental attributes of the empty PRF against those of known roosts.

This chapter describes the values that we need to record for comparison, and means that at the end of a day where no bats have been encountered, you will nevertheless have a useful body of data that will, if submitted to the BRHK project, mean that knowledge of the sort of PRF different rock types can offer will build over time to a point where a reliable confidence threshold is identified for the benefit of: **a)** the bats themselves; **b)** the surveyor; and, **c)** any quarry developer that wants to leave behind a rock face that will have a tangible value to bats as part of a restoration design.

Finally, this chapter sets out the evidence that might be used to assess whether or not the empty PRF is suitable for a particular bats species to exploit it as a roost, and whether it is in fact a roost already.

3.2 What

The 'what' is all the meso- and micro-scale environmental values.

3.2.1 The recording form

To refresh your memory, the *BRHK Rock Face Database* recording form is provided again in Table 3.1.

> **Note:** Although some categorical variables have the options provided, they are prompts and not restrictions. What that means is that if a single landform or PRF encompasses many variables they should all be circled.

The values on the top half of the form comprise a basic landform and PRF record, and encompass the meso-environment variables that characterise the situation and form of the PRF, and of course the situation any bat(s) might be roosting in on the day. This can be considered the 'mapping record' too, because with an annotated image it would allow someone else to return to the habitat, find the PRF on the rock face and reinspect it.[9]

The bottom half of the form may be considered the 'advanced record'. These are the micro-environment variables.

3.2.2 Introducing the rock face meso-environment values

The meso-environment is what a bat encounters as it flies through the landscape and encompasses the landform rock type, the landform itself, and the situation in which the bat might roost.

In the context of the BRHK project the rock face meso-environment values comprise: **1)** the landform rock type; **2)** the landform; **3)** the habitat around, on, and over the landform and into which the PRF opens; **4)** the rock face height; **5)** the rock face aspect; **6)** the rock face angle; **7)** the PRF type; **8)** the PRF height; **9)** the PRF entrance aspect; **10)** the PRF entrance angle; **11)** the entrance height; and, **12)** the entrance width.

3.2.3 Introducing the rock face roost account

The evidence of roosting is placed between the meso-environment and the micro-environment, because the meso-environment coupled with roosting evidence comprises the basic record of the roost. This information is sufficient to inform searches for other similar roosts. However, if the recorder is really interested in contributing to a predictive framework for survey data analysis, and a detailed design specification for making artificial roosts elsewhere, they might like to expand their recording to include the micro-environment values.

3.2.4 Introducing the rock face micro-environment values

The micro-environment encompasses the internal characteristics of the PRF and are what the bat experiences. In the context of the BRHK project the micro-environment values comprise: **1)** the internal height (i.e. how far above the entrance the feature extends upward); **2)** the internal width (i.e. how far into the rock and away from the entrance the PRF extends in parallel to the ground); **3)** the internal depth (i.e. how far below the entrance the feature extends downwards); **4)** the apex shape; **5)** the internal substrate; **6)** the cleanliness inside; **7)** the humidity (in broad terms); **8)** whether any competitors also occupy the PRF (i.e. any other species with which a bat would have to share the PRF); and, **9)** whether comprehensive inspection was possible.

Table 3.1 The *BRHK Rock Face Database* recording form.

BASIC RECORD	**WHO**	**1st Recorder:**		**2nd Recorder:**
	WHERE	**Site name:** see Chapter 1		
		Grid reference: see Chapter 1		
	WHEN	**Date:** see Chapter 1		
	WHAT	Rock type (e.g. granite):		
		Is the land form Landform in: an escarpment / a crag / a tor / a gorge / a boulder or block / a cutting / a quarry		
		Habitat on/over the landform (phase 1):		
		Rock face height (m):		
		Rock face aspect (e.g. North):		
		Is the rock face angle: reclined / sheer / overhung		
		What is the bat occupying: a crack / a break / a flake / a mono / a bidoigt / a pocket / a bucket / an off-width		
		PRF height (measured from base of face) (m):		
		PRF entrance aspect (e.g. North):		
		PRF entrance angle (e.g. 90° use protractor):		
		PRF entrance height (m / cm):		
		PRF entrance width (m / cm):		
ADVANCED RECORD	**WHAT**	**ROOST ACCOUNT**	Bat species:	
			Number of bats:	
			Roost position in relation to the entrance: above / in front / to the side / below	
			Distance from entrance (cm):	
			Awake or torpid:	
			Droppings:	
			Odour cues (i.e. smell): none / pleasant / not unpleasant / repellent	
		PRF INTERNAL	Height (i.e. top of entrance to top of interior) (cm):	
			Width (i.e. inside entrance lip to the back wall in front) (cm):	
			Depth (i.e. bottom of entrance to bottom of interior) (cm):	
			Apex shape: a spire / a peak or wedge / flat / unknown	
		INTERNAL CONDITIONS Circle as many of the fields as are applicable	**Internal substrate:** glossy smooth / flowing bumpy / sandpaper rough / jagged	
			Internal cleanliness:	*Primary* – clean / waxy / stained / polished
				Secondary – messy / dusty with loose debris / muddy or silty / sludgy
			Internal humidity: dry / damp / wet / green algae	
			Competitors:	
COMPREHENSIVE INSPECTION POSSIBLE (yes or no):				

Note: This is a lot of information to record at the outset, but repeat inspections will effectively only have to record whether bats were there or not, and then the four values that deal with the substrate, cleanliness, humidity and competitors, because these are the ones that may change between inspections.

The following sections describe the individual values so the surveyor can recognise them and record them.

Note: The following text refers to five sources of roost records, as follows:

1. The *BRHK Rock Face Database* records submitted by teams in Yorkshire and Cardiff as follows:

   The Yorkshire team: Robert Bell (team leader), Brian Armstrong, Dan Best, Andrew Hill, Jamie Ingram, Paul Liptrot, Amanda Murphy, Scott Reed, Greg Slack, Katie Smith, Dan Wildsmith, Steven Whitcher and Ian Wright. This team has recorded and photographed roosts of: 1) whiskered bats *Myotis mystacinus*; 2) Natterer's bats *M. nattereri*; 3) common pipistrelles *Pipistrellus pipistrellus*; 4) soprano pipistrelles *Pipistrellus pygmaeus*; and, 5) brown long-eared bats *Plecotus auritus*.

   The Cardiff team: Hal Starkie (team leader), Jess Dangerfield, James Humphries, James Shipman, Aaron Davies, Pippa Loam, and Megan Brenchley. This team has recorded and photographed roosts of: 1) serotines *Eptesicus serotinus*; 2) Daubenton's bats *Myotis daubentonii*; 3) Natterer's bats; and, 4) brown long-eared bats.

2. Records made by Jean Matthews of: 1) a barbastelle *Barbastella barbastellus* roost in France; and, 2) a Natterer's bat roost in Conwy.

3. Two serotine roosts recorded in Wales by Sam Dyer (Dyer 2013; also S. Dyer, pers. comm., June 2013).

4. A record of a Daubenton's bat roost in Norway provided by Rune Sørås (R. Sørås, pers. comm., October 2019).

5. A brown long-eared bat roost recorded in Leicestershire by Nathalie Cossa.

Where data were not recorded in the field but sufficient information is present in photographic evidence to use in an account or analysis, the data are identified as estimated from a photograph with *

Finally, one report source is referred to, comprising:

- Shewring, M. 2021. Evidence of rock exposure roosting bat species in the UK from recreational rock climbers. Unpublished manuscript, Cardiff University.

Unless otherwise referenced, all data relate to these resources.

3.3 What 1: recording the rock face meso-environment values

3.3.1 Rock type

Thus far, the limited data in respect of which rock types hold PRF on rock faces that are exploited by roosting bats are summarised in Table 3.2. This includes confirmed roosts and anecdotal reports.

Those people who are setting out to record the PRF in a specific face might identify the rock in advance using the British Geological Survey website.[10] In order to save the rock climbers some time, we have compiled an inventory of climbing sites with the rocks the faces comprise and this is provided at Appendix B. To clarify, the reason the rock is important is that different rocks may form different PRF and offer different environments due to factors such as grain, porosity and heat retention. These environments may be exploited in different ways in different seasons by different bat species.

By cataloguing the rocks, particular trends may be identified for use in a predictive analysis. This will allow the recorder to mentally prepare themselves to search for specific PRF types in different situations.

3.3.2 Landform

To recap on Chapter 2 (which had initial photographic illustrations for reference), the rock face landforms encompass:

» **Escarpments** – Escarpments comprise a long and sheer cliff separating flat land at different levels. The plateau above and the land at the foot of the cliff are broadly level or have a gentle slope. In the context of this project, escarpments include sea-cliffs.

» **Crags** – Crags encompass all vegetated rock promontories that lead into a tail of rock fragments and soil.

» **Tors** – Tors encompass all bare rock promontories that may or may not lead into a tail of rock fragments.

Table 3.2 The rock types within which roosting bats have been recorded or reported occupying PRF, in rock faces.

ROCK TYPE	Record held on the *BRHK Rock Face Database*	Record in white paper or photographic account	Anecdotal account reported
Mudstone	—	Ancillotto *et al.* (2014) – N.B. study performed in Italy	—
Dolerite	—	—	Shewring (2021)
Gneiss	✓	—	Shewring (2021)
Granite	—	—	Shewring (2021)
Greywacke	—	—	Shewring (2021)
Gritstone	✓	—	Shewring (2021)
Limestone	✓	Dyer (2013)	Shewring (2021)
Rhyolite	—	—	Shewring (2021)
Sandstone	✓	—	Shewring (2021)
Schist	✓	—	Shewring (2021)
Tuff	✓	—	—

» **Gorges** – Gorges encompass all narrow valleys between hills or mountains including dry 'fossil' ravines and those with rivers or streams in the base.

» **Boulder/block** – Any massive, rounded boulder or angular block that is too small to be a tor but large enough to represent an immovable landmark. Boulders/blocks may occur as isolated individuals (e.g. glacial erratics)[11] or in talus and boulderfields.[12]

» **Cuttings** – Cuttings encompass all passages cut through an obstacle of rock to accommodate a railway line, road or any other route. Some cuttings are effectively artificial gorges, others are simply one-faced and cut into rock in order to bypass another topographical obstacle.

» **Quarries** – A quarry is an open pit from which stone or ore is won for use in construction, manufacturing and even pharmaceuticals. A single quarry may encompass artificial escarpments, crags, tors and even gorges, all in miniature. It is not uncommon for such a quarry to also encompass, some or all of the following: artificial scree, talus, natural caves, mine entrances, and tunnels.

The *BRHK Rock Face Database* records and the photographic reports are summarised in Table 3.3.

The data are tantalisingly incomplete and there is room for someone to make a name for themselves by being the first to record a bat of any species in a tor or boulder/block, and by being the first to record additional bat species in all of them.[13]

3.3.3 Habitat on/over the rock landform

This relates to whether or not the landform itself is vegetated, and if so, what the vegetation comprises. It also relates to the habitat surrounding the landform. It may be that an individual bat species tolerates and even favours open and exposed situations generally or in specific periods of the year and not at other times. It may be that some

Table 3.3 Rock Face landforms that are proven to be exploited by individual bat species.

BAT SPECIES	LANDFORM						
	NATURAL					ARTIFICIAL	
	Escarpment	Crag	Tor	Gorge	Boulder/block	Cutting	Quarry
Barbastelle *Barbastella barbastellus*	?	✓	?	?	?	?	?
Serotine *Eptesicus serotinus*	✓	?	?	✓	?	?	✓
Daubenton's bat *Myotis daubentonii*	?	?	?	✓	?	?	?
Whiskered bat *Myotis mystacinus*	?	?	?	✓	?	?	✓
Natterer's bat *Myotis nattereri*	?	✓	?	✓	?	✓	✓
Common pipistrelle *Pipistrellus pipistrellus*	✓	✓	?	?	?	?	✓
Soprano pipistrelle *Pipistrellus pygmaeus*	?	?	?	?	?	✓	?
Brown long-eared bat *Plecotus auritus*	?	✓	?	✓	?	✓	✓

bat species favour landforms that are under woodland cover and tolerate the cooler conditions. It may be that flooded quarries are favoured over dry quarries, or conversely that dry 'fossil' gorges are favoured over active riparian situations.

It may be that different species timeshare a specific roost feature on a specific landform over the annual cycle and tolerate/favour different environments at different times of year. The habitat is recorded using the 'Handbook for Phase 1 habitat survey – a technique for environmental audit' (JNCC 2010) which can be downloaded free of charge.[14] However, to save some time at the start, the rock landforms fit into the following Phase 1 habitats:

» **Escarpment** – I1.1 – Rock exposure and waste / Natural / Inland cliff **or** H8.1 – Coastland / Maritime cliff and slope / Hard cliff;

» **Crag** – I1.1 – Rock exposure and waste / Natural / Inland cliff;

» **Tor** – I1.1 – Rock exposure and waste / Natural / Inland cliff;

» **Gorge** – I1.1 – Rock exposure and waste / Natural / Inland cliff;

» **Boulder/block** – An individual boulder or block would rationally be Target Noted, but where they occur in a boulderfield that Target Note would be within the wider habitat classification of I1.4 – Rock exposure and waste / Natural / Other exposure;

» **Cutting** – J5 – Miscellaneous / Other habitat;

» **Quarry** – I2.1 – Rock exposure and waste / Artificial / Quarry.

'I1.4 – Rock exposure and waste / Natural / Other exposure' is a useful catch-all in rocky areas, **but that is missing the point**; in fact, the idea of this section is to attempt to understand whether vegetation generally (and woodland in particular) can be used to predict the presence of a specific bat species at a specific point in the year.

Of the 42 records on the *BRHK Rock Face Database*, and the six photographic records, 45 records are associated with A1.1.1 – Broadleaved semi-natural woodland, and three are associated with D1.1 – Dry dwarf-shrub heath. However, at present we have no replicate data from exposed situations, so all this can tell us is that bats do use vegetated situations, not that they avoid all other situations.

> **Note:** Although it is labouring the point, if we have the full 12-value alphanumeric OS grid reference, we can look at the habitat remotely using satellite images to see if there is anything common to roost situations occupied by specific bat species, and anything that separates them.

3.3.4 Rock face height

The rock face height is the height of the individual continuous face in which the PRF is situated.

While escarpments, boulders/blocks and cuttings typically comprise a single continuous face from the foot to the summit, crags, tors and gorges commonly encompass more than one face and these may be complex in their arrangement. Similarly, a quarry face may be tiered in stepped benches.

The rule is simple: the face is measured from surveyor's feet when they can stand at the foot of the face and go no lower. The top is the summit where there is nothing more to climb. The measurement is taken in metres. Accuracy may be to 0.5 of a metre if the surveyor is confident in their equipment, in most cases to the nearest metre will be fine.

3.3.5 Rock face aspect

The aspect of a rock face will determine how exposed it is to the sun's rays and therefore how hot it gets in sunny weather and also how much light hits it, when and for how long each day. This meso-environment aspect may thus influence the micro-environment temperature, humidity and illuminance gradients. It will also influence when the face is sufficiently dark for bats to begin emerging. The aspect of the rock face is recorded using a compass and to the nearest eighth, e.g. NE.

As the reliability of the reports is unknown, they were not combined with the Database and photographic records to give one overarching summary, but instead are considered separately.

Records held on the *BRHK Rock Face Database* are summarised as follows:

- » **North** – 13% (n=6);
  - – North-east – 2% (n=1);
- » **East** – 4% (n=2);
  - – South-east – 6% (n=3);
- » **South** – 11% (n=5);
  - – South-west – 54% (n=26);
- » **West** – 4% (n=2);
  - – North-west – 6% (n=3).

Aggregating south-east, south, and south-west catches 71% of the records within 25% of the compass.

Looking separately at the unsupported anecdotal reports submitted by rock climbers and collated by Shewring (2021), the data are presented in the same format as the records, as follows:

- » **North** – 4% (n=2);
  - – North-east – Nil;
- » **East** – 8% (n=4);
  - – South-east – 8% (n=4);
- » **South** – 35% (n=17);
  - – South-west – 10% (n=5);
- » **West** – 31% (n=15);
  - – North-west – 4% (n=2).

Aggregating south-east, south, and south-west catches 53% of the reports within 25% of the compass.

While it is accepted that the exercise above is data mining, and it cannot be overemphasised that roosts have been recorded and reported in faces with northerly, easterly and westerly aspects, the data do appear to suggest that there is a greater probability in encountering a bat in face with some degree of southerly aspect.

3.3.6 Rock face angle

The rock face angles are categorised as follows:

- » **Reclined** – The cliff section is a slope that might be scrambled up, and if someone fell down it they would roll or slide;
- » **Sheer** – The cliff section is broadly vertical and could not be scrambled up, and if someone fell down it (while they might bounce off any outcrops) their descent would be a fall;

» **Overhung** – The top of the cliff section is leaning out further than the base and if anyone fell down it their descent would be a plummet with nothing to break their fall between the lip and the ground at the foot.

Note: Although the photographs in Chapter 1 illustrate beautifully simple situations, the larger the face the more angles it may encompass. This is particularly common in old quarries where it is not unusual to find areas of a single continuous face that encompass sections of all three angles, and even vertically from the foot to the top, and not in a convenient order. For example, a cliff that is reclined at the foot (typically where the face has been back-filled) is sheer on the midsection and then juts out with an overhang above. Figure 3.1 illustrates combinations of angles.

Top to bottom:

Overhung top

Sheer midsection

Reclined foot

Figure 3.1 Combinations of angles on a limestone quarry face.

3.3.7 PRF type

The PRF types encompass both crevices and voids.

Crevice-type PRF are classified by their orientation which makes them either: **a)** a crack; **b)** a break; or **c)** a flake.

The specific criteria are as follows:

» **Cracks** – Vertical crevices that extend perpendicular into the cliff face;
» **Breaks** – Horizontal crevices that extend perpendicular into the cliff face;
» **Flakes** – Vertical crevices comprising a detached slab of the cliff face. Flakes may open top down, bottom up (the base of the flake providing an overhang), or flanking across the face. However, one factor links them all: the entrance is in a corner. If the entrance is perpendicular to the surface on either side, it is a crack and not a flake.

Figures 3.2 and 3.3 illustrate the context of crack PRF and provide close-up images of individual examples.

Figure 3.2 Crack PRF in the context of the overall face (in this case limestone).

Figure 3.3 Examples of individual crack PRF (all limestone).

Figures 3.4 and 3.5 illustrate the context of break PRF and provide close-up images of individual examples.

Figure 3.4 Break PRF in the context of the overall face (in this case sandstone). N.B. The landform in this image is in fact all that remains of a broken glacial gorge-side. It might be identified as a tor, and demonstrates the need to 'interpret' the landform on the context of the landscape, rather than simply 'identify' in isolation.

Figure 3.5 Examples of individual break PRF. Left: limestone. Middle: sandstone. Right: granite.

Figures 3.6 and 3.7 illustrate the context of flake PRF and provide close-up images of individual examples.

Figure 3.6 Flake PRF in the simplest context.

Figure 3.7 Examples of individual flake PRF. Left: limestone. Middle: slate. Right: limestone.

Note: When faced with a diagonal crevice in the field there can be some doubt as to at what point between 90° and 0° a crack becomes a break. From a technical perspective anything 46° to 90° might be thought of as a crack and anything from 0° to 45° a break, but in fact that is not functionally accurate. In rock climbing a crack is not as helpful in upward momentum as a break: this is logical, as a climber reaching up to a horizontal break can use it to pull themselves up, whereas cracks are better for sideways movement and to hold the climber to the face. The litmus test is therefore to ask, 'if I put both my hands in this crevice, could I securely hang with my feet off the ground or pull myself upward?' If the answer is yes then it is a break. If the answer is no then you may quite reasonably classify the crevice as a crack.

In addition, to remove any nagging doubts as to whether you have a crack or a flake, Figure 3.8 illustrates the difference.

The feature on the left is in a flat face and is perpendicular to the surface: this is a <u>crack</u>.

The feature on the right is in a corner, and is flanking (parallel) to the face: this is a <u>flake</u>.

Figure 3.8 The difference between a crack and a flake (both in limestone).

Finally, we come to the void-type PRF. These have also been divided by climbers into:

» **Mono** – A hole that is only big enough for one finger;
» **Bidoigt** – A hole that has enough room for two fingers;
» **Pocket** – A hole in the cliff face that is big enough and extends in far enough to be used as a hand-hold;
» **Bucket** – A hole bigger than a hand but not larger than typical bucket-size;
» **Off-width** – An alcove opening into the face. N.B. Anything larger is entering the dimensions of a chamber or cave and would no longer be classified as a cliff PRF.

Individual mono, bidoigt, pocket and bucket PRF are easily visualised, but an off-width is less so and one is therefore illustrated in Figure 3.9.

Figure 3.9 An example of an off-width.

Table 3.4 PRF Types for which there is evidence of exploitation by specific bat species.

BAT SPECIES	PRF TYPE							
	CREVICE-TYPE		VOID-TYPE					
	Crack	Break	Flake	Mono	Bidoigt	Pocket	Bucket	Off-width
Barbastelle *Barbastella barbastellus*	?	✓	?	?	?	?	?	?
Serotine *Eptesicus serotinus*	✓	?	✓	?	✓	?	?	?
Daubenton's bat *Myotis daubentonii*	✓	✓	?	?	?	?	?	?
Whiskered bat *Myotis mystacinus*	✓	✓	✓	?	?	?	?	?
Natterer's bat *Myotis nattereri*	✓	✓	✓	?	?	?	?	✓
Common pipistrelle *Pipistrellus pipistrellus*	✓	✓	✓	?	?	?	?	?
Soprano pipistrelle *Pipistrellus pygmaeus*	✓	?	?	?	?	?	?	?
Brown long-eared bat *Plecotus auritus*	✓	✓	✓	?	✓	?	?	?

Note: In most natural weathered faces the PRF are typically discrete and clearly defined as one specific feature that can be comprehensively inspected. In quarries, and limestone quarries in particular, the PRF are typically interconnected in a network of fractures, and a single face may be a mass of every type of PRF, in a multitude of dimensions. Hence, the face shown in Figure 3.10 is by no means unusual.

A fractured limestone face of monos, bidoigts, pockets and buckets

Figure 3.10   A quarry face of monos, bidoigts, pockets and buckets.

The reader will doubtless have noted that many of the illustrative images used so far comprise artificial rock faces in quarries. The reason for this is twofold: **1)** they are the faces most professional ecologists are typically asked to survey; and, **2)** they are significantly more complex than a typical natural rock face – if you can map a quarry face, you can map pretty much anything!

The photographic records we have seen are coupled with the records on the *BRHK Rock Face Database,* to give a summary of PRF types we can conclusively state are exploited in Table 3.4.

3.3.8 PRF height

The PRF height is taken to be the height in the centre of the entrance into the PRF measured from the foot of the rock face.

In some situations, there may be doubt as to where the foot of the face is. This is because some faces are stepped, e.g. quarry faces that have benches, railway lines that have been cut into the side of a gorge, or simply an exposed face that is some way up the side of an otherwise soil-covered vegetated slope.

In all situations, the basic principle is that the height is taken from the surveyors' feet in the position that they could comfortably stand to take the measurement.

At the moment, as with the *Bat Tree Habitat Key (BTHK) Database* for tree roosts, the *BRHK Rock Face Database* is biased to ground inspections. Notwithstanding, the summary provided in Table 3.5 is the best available evidence.

3.3.9 PRF entrance aspect

The surfaces of rock faces may be complex in themselves due to the grain of the rock. In addition, sedimentary, biogenic and metamorphic rocks have strata that may be orientated in several directions over a single landform. As a result, not all the PRF on an individual rock face may be orientated outward and perpendicular to the face.

In addition to choosing a roost with a comfortable stable temperature, bats may prefer a position where they would themselves perceive daylight but not be visible if the sun was directly on the roost entrance, and this may have a bearing upon whether or not PRF with a specific entrance aspect are occupied. As with face aspect, PRF aspect is defined by the points of the compass to the nearest eighth, e.g. NE.

3.3.10 PRF entrance angle

The angle of the entrance is recorded using a protractor parallel to the face.

Angles are to a certain extent integral to the PRF type: cracks and flakes denote broadly vertical orientation where a break denotes broadly horizontal orientation. The data are

Table 3.5 Height ranges of PRF occupied by bats in rock faces.

BAT SPECIES	HEIGHT RANGE (m)
Barbastelle *Barbastella barbastellus*	1.5*
Serotine *Eptesicus serotinus*	1.45–20*
Daubenton's bat *Myotis daubentonii*	2.2–20*
Whiskered bat *Myotis mystacinus*	1.48–5
Natterer's bat *Myotis nattereri*	0.8–6
Common pipistrelle *Pipistrellus pipistrellus*	1–9
Soprano pipistrelle *Pipistrellus pygmaeus*	3
Brown long-eared bat *Plecotus auritus*	0.75–5

*Value estimated from photographic record.

as yet insufficient to be usefully summarised, but continued recording of the angle to within 10° may identify partitioning in different species and different seasons.

This can be recorded in the field or by reference to photographic images when the surveyor returns to the office.

3.3.11 PRF entrance height and width

The entrance height is recorded from the base to the apex to give a representative value. The measurement is taken to the nearest 0.5 cm.

> **Note:** Some people prefer recording in mm. This is fine when measurements are typically less than 1 cm, but when the measurements get above 10 cm problems creep in. Our advice is to be sensible: centimetres are a really useful tool. Unless you are a physicist or an engineer, we recommend using centimetres whenever you are recording a value that is under a metre and over 9 mm.
>
> However, we recognise that most rock climbers will not be carrying a tape measure, and the standard climbing measurements may be used:
>
> - **Seam** – A crevice too narrow to take fingers (< 2 cm);
> - **Finger** – A crevice wide enough to take fingers, but only up to the knuckles (2–3 cm);
> - **Hand** – A crevice wide enough to place the whole hand flat or slightly cupped (3.1–4.5 cm);
> - **Fist** – A crevice that is wide enough to clench a fist in order to use it as a jamming hold (> 4.5 cm).

For the removal of doubt, in a crack or the greater proportion of flakes the entrance height will be a significantly greater value than the entrance width. In the case of a break, the entrance height will be significantly less than the entrance width.

The PRF entrance width is recorded across the horizontal. When faced with a long crack that has undulating sides it can be difficult to decide where to take the measurement. Where this happens, take two measurements: the first is the general width across the entire length, the second is the maximum. We present these as follows:

$$\text{general (maximum)}$$

$$\text{e.g. 2 cm (11 cm)}$$

These values are important when we consider that several bat species appear to favour crevice-roosts where they are against a substrate on both their dorsal (back) and ventral (chest) surfaces, but other species appear to favour a more open situation where they roost with only their ventral surface against the substrate and their back exposed, and some others appear to favour situations where they hang entirely free of the substrate. Obviously, these are generalisations and there are individual bats that like to buck the trend, but these tend to be the exceptions that prove the rule. Table 3.6 summarises the range of entrance dimensions of roosts thus far recorded on the *BRHK Rock Face Database* and estimated from photographic records.

Table 3.6 A summary of the range of entrance dimensions of roosts thus far recorded on the *BRHK Rock Face Database* and estimated from photographic records. Values given are centimetres. Where one value is given this reflects the fact that there is only one pertinent record. Where are pair of values are given these are the minimum and maximum.

| BAT SPECIES | ENTRANCE HEIGHT & WIDTH RANGE (cm) | PRF TYPE | | | | | | | |
| | | Crevice-type | | | Void-type | | | | |
		Crack	Break	Flake	Mono	Bidoigt	Pocket	Bucket	Off-width
Barbastelle *Barbastella barbastellus*	Height >>	?	2	?	?	?	?	?	?
	Width >>	?	20	?	?	?	?	?	?
Serotine *Eptesicus serotinus*	Height >>	33	?	30*	?	5	?	?	?
	Width >>	2–6	?	4*	?	5	?	?	?
Daubenton's bat *Myotis daubentonii*	Height >>	50*	2	?	?	?	?	?	?
	Width >>	2*	63	?	?	?	?	?	?
Whiskered bat *Myotis mystacinus*	Height >>	32	1.5	1.5	?	?	?	?	?
	Width >>	4	40	50	?	?	?	?	?
Natterer's bat *Myotis nattereri*	Height >>	25–80	5	2	?	?	?	?	18
	Width >>	2	?	25	?	?	?	?	5
Common pipistrelle *Pipistrellus pipistrellus*	Height >>	20–1,500	1.5–6	30–150	?	?	?	?	?
	Width >>	1–10	32–250	1.5–2.5	?	?	?	?	?
Soprano pipistrelle *Pipistrellus pygmaeus*	Height >>	15	?	?	?	?	?	?	?
	Width >>	2	?	?	?	?	?	?	?
Brown long-eared bat *Plecotus auritus*	Height >>	3–300	?	8–48	?	?	?	?	?
	Width >>	2–28	?	1.5–2	?	?	?	?	?

* Value estimated from photographic record.

Note: It is not uncommon for the entrance to a PRF to go round a corner or even be an S/Z shape: the entrance is the full length of the crevice from one end of the opening to the other as in the example shown in Figure 3.11.

The entrance to a complex PRF such as this flake is recorded over the full length, corners and all

Figure 3.11 Elongate complex entrance height.

Table 3.7 The range of entrance dimensions occupied by all species of bats, and those more likely than not to be exploited.

PRF TYPE	ENTRANCE AXIS	OVERALL RANGE (cm)	GREATER THAN 50% (cm)
CRACK	Height >>	15–1,500	35.5–177.5 (n=26)
	Width >>	1–10	2–3.3 (n=27)
BREAK	Height >>	1.5–6	1.8–2.5 (n=9)
	Width >>	15–250	28–63 (n=9)
FLAKE	Height >>	1.5–150	8–48 (n=9)
	Width >>	1.5–50	1.8–4 (n=9)
BIDOIGT	Height >>	5	5 (n=1)
	Width >>	5	5 (n=1)
OFF-WIDTH	Height >>	18	5 (n=1)
	Width >>	18	5 (n=1)

To give a useful general tool to inform triage in the field, the entrance dimensions of roosts occupied by all species were calculated and are provided from the perspective of the PRF type. The results are provided in Table 3.7, which shows the full range and also the dimensions that greater than 50% of the roosts exhibited.

3.4 What 2: recording the rock face roost account

3.4.1 The bats

Although the data with sufficient detail to be of practical value in the field comprise just eight bat species, an overall 12 bat species that occur in the British Isles are reported to exploit features in rock-face habitat as roost sites, these comprise:

1. **Barbastelle** *Barbastella barbastellus* (Stebbings 1988; Ancillotto *et al.* 2014; *BRHK Rock Face Database*);
2. **Serotine** *Eptesicus serotinus* (Stebbings 1988; *BRHK Rock Face Database*);
3. **Daubenton's bat** *Myotis daubentonii* (Barrett-Hamilton 1910; Altringham 2003; Michaelsen *et al.* 2013; R. Sørås, pers. comm., October 2019; *BRHK Rock Face Database*);
4. **Whiskered bat** *Myotis mystacinus* (Michaelsen *et al.* 2013; *BRHK Rock Face Database*);
5. **Natterer's bat** *Myotis nattereri* (*BRHK Rock Face Database*);
6. **Noctule** *Nyctalus noctula* (Harrison 1962; Schober & Grimmberger 1993; Stebbings 1988);
7. **Leisler's bat** *Nyctalus leisleri* (Stebbings 1988; Shiel *et al.* 2008);
8. **Nathusius' pipistrelle** *Pipistrellus nathusii* (Dietz *et al.* 2011);
9. **Common pipistrelle** *Pipistrellus pipistrellus* (Vesey-Fitzgerald 1949; Schober & Grimmberger 1993, 1997; Altringham 2003; *BRHK Rock Face Database*);
10. **Soprano pipistrelle** *Pipistrellus pygmaeus* (Vesey-Fitzgerald 1949; Schober & Grimmberger 1997; Altringham 2003; *BRHK Rock Face Database*);

11. **Brown long-eared bat** *Plecotus auritus* (Whitaker 1906; Stebbings 1988; *BRHK Rock Face Database*);

12. **Grey long-eared bat** *Plecotus austriacus* (Stebbings 1988).

The four bat species for which the BRHK project has no data comprise: **1)** noctule; **2)** Leisler's bat; **3)** Nathusius' pipistrelle; and, **4)** grey long-eared bat.

This project is all about standardisation of records, providing proof that the records are accurate, and gathering information that will have a practical application in informing searches and producing artificial roost designs. To the latter end, the values recorded when bats are present comprise: **a)** the species; **b)** the number of bats; **c)** whether the bats are above the entrance (i.e. they have gone in and up), to one side (i.e. they have gone in and to one side or just in and straight back) or below the entrance (i.e. they have gone in and down); **d)** the distance the bats are from the entrance; and, **e)** whether the bats are awake or asleep/torpid (i.e. are they obviously active or simply have their eyes open, or are they entirely static and have their eyes closed).

However, before we rush on it is worth identifying 'when' bats have been recorded in rock faces. The summary is provided below:

» **Winter (January/February)** – All species 39% (n=35);
  - Serotine – 5% (n=4);
  - Daubenton's bat – 1% (n=1);
  - Whiskered bat – 1% (n=1);
  - Natterer's bat – 2% (n=2);
  - Common pipistrelle – 16% (n=14);
  - Soprano pipistrelle – 1% (n=1);
  - Brown long-eared bat – 13% (n=12).

» **Spring-flux (March/April)** – All species 19% (n=17);
  - Serotine – 8% (n=7);
  - Common pipistrelle – 8% (n=7);
  - Brown long-eared bat – 3% (n=3).

» **Pregnancy (May/June)** – All species 11% (n=10);
  - Barbastelle – 1% (n=1);
  - Serotine – 1% (n=1);
  - Whiskered bat – 2% (n=2);
  - Natterer's bat – 2% (n=2);
  - Common pipistrelle – 4% (n=3);
  - Brown long-eared bat – 1% (n=1).

» **Nursery (July/August)** – All species 7% (n=6);
  - Serotine – 4% (n=3);
  - Daubenton's bat – 1% (n=1);
  - Common pipistrelle – 2% (n=2).

» **Mating (September/October)** – All species 4% (n=3);
  - Common pipistrelle – 4% (n=3);

» **Autumn-flux (November/December)** – All species 20% (n=18);
  - Natterer's bat – 2% (n=2);
  - Common pipistrelle – 12% (n=11);
  - Soprano pipistrelle – 1% (n=1);
  - Brown long-eared bat – 5% (n=4).

A summary of the data aggregated for all bat species is illustrated in Figure 3.12.

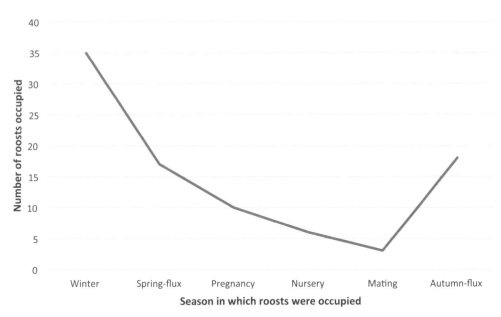

Figure 3.12 The number of records of bat roosts in rock faces occupied in each season of the year (data taken from *BRHK Rock Face Database* and photographic records).

Despite the bias towards occupancy in the cooler seasons, the bats occupying rock faces generally appear to be in relatively exposed situations, as per the examples on the book cover and the records held on the *BRHK Rock Face Database* presented in Table 3.8.

Table 3.8 A summary of roost position data held on the *BRHK Rock Face Database*.

BAT SPECIES	% of roost positions above the entrance with minimum & maximum distance (cm)	% of roost positions in front or to one side of the entrance with minimum & maximum distance (cm)	% of roost positions below the entrance with minimum & maximum distance (cm)
Barbastelle *Barbastella barbastellus*	—	100% 5*	—
Serotine *Eptesicus serotinus*	—	100% 5–25	—
Daubenton's bat *Myotis daubentonii*	—	100% 5–13	—
Whiskered bat *Myotis mystacinus*	67% 10–20	33% 6	—
Natterer's bat *Myotis nattereri*	50% 10–20	50% 5–10	—
Common pipistrelle *Pipistrellus pipistrellus*	29% 5–20	71% 2–32	—
Soprano pipistrelle *Pipistrellus pygmaeus*	—	100% 5–10	—
Brown long-eared bat *Plecotus auritus*	29% 3–25	71% 3–35	—

* Value estimated from photographic record.

To summarise the data in Table 3.8, all the bats roosting above the entrance were no more than 25 cm away from the outside air, and those roosting in front or to the side of the entrance were no more than 35 cm away.

When the bat is in such an exposed situation, identification may be straightforward. However, where it is not then we would urge that where possible a photograph is taken side-on and then a photo of the bat's face is taken using an endoscope. If you are using a 17 mm lens either dim the lamp or drop it to a 9 mm because the lighting is less disturbing when going for a close-up. The photograph is to capture the ear shape, and the muzzle, this is usually sufficient to identify the bat to genus (family) level. If you can get a photograph that illustrates the tragus length and shape, or the pelage (without agitating the bat), then this will aid an identification to species. However, bear in mind that you are not going for an award-winning image, just something sufficient to inform an identification.

In most cases, a photograph can be uploaded to the BRHK Facebook page and will at least result in a confident genus-level identification. Species-level gaps are 'ironed out' in the data over time so a failure to get to species is not the end of the world. In all cases, where droppings are present, we would urge that these are taken for DNA analysis rather than a torpid bat subjected to prolonged illumination.

3.4.2 Droppings

Where possible, it is wise for the 1st Recorder to use a torch first and look in the base of the PRF before putting the endoscope in. This is because inspections using an endoscope have the potential to dislodge and crush individual droppings. The presence of droppings focuses attention and may give some indication of whereabouts inside the PRF the bats are hung up and this will inform any subsequent endoscope inspection. Obviously, anyone using an endoscope will take care throughout every inspection, but even the most experienced user will benefit from having some idea of the orientation of the bats before they begin.

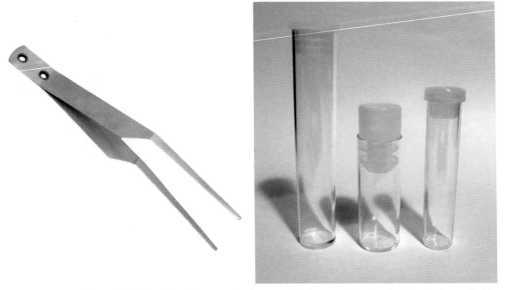

Figure 3.13 Left: storkbill fine blunt forceps. Right: a push-top specimen tube. Both are available from nhbs.com.

It is sensible to take a dropping sample using forceps and place it into a sample tube. We favour the storkbill fine blunt forceps sold by NHBS because they do not take chunks out of us when we are trying to get them out of a bag one-handed while hanging in a precarious position. For the same reason, pull and push-top specimen tubes are better than screw-tops. Both blunt forceps and push-top specimen tubes are illustrated in Figure 3.13.

If no bats are present on the day of the inspection, or bats are present but the identification is uncertain, the droppings may be submitted to Warwick University for DNA analysis. If bats are recorded and a confident identification is achieved, then the droppings can simply be discarded.

3.4.3 Odour cues

Odour cues have been found to be a useful field-sign in determining the roosts status of PRF in trees (see BTHK 2018), but thus far odour does not appear to be so useful with roosts in rock. As there is the possibility that the odour perceived in trees is a chemical combination of both bat and wood, the decision was made not to include the tree-associated odours in this book. Notwithstanding, if the surveyor wishes to attempt to record the odour of the roost the categories comprise:

» **None** – no odour detectable;

» **Pleasant** – a smell the person experiencing finds pleasant and attractive;

» **Not unpleasant** – a smell that is neither attractive or repellent but simply tolerable;

» **Repellent** – a smell that is something the surveyor would rather not be exposed to ever again.

3.5 What 3: recording the rock face micro-environment values

The macro- and meso-environment combine to deliver part of the micro-environment, the rock and competitors deliver another part, but the bats themselves deliver the rest by their own body-heat, breath and movement inside the roost.

The micro-environment values may be thought of as the 'advanced record'. These are the values that characterises the environment the bats actually experience and influence inside the roost, and they encompass: **a)** the dimensions and physical shape of the space the bats occupy; **b)** the substrate over which the bats must pass and on which they will either rest against or hang from; **c)** the cleanliness of the situation; **d)** the humidity; and, **e)** any competitors that may also want to live in the PRF with or without the bats.

In addition to the presence of the bats themselves, the micro-environment values that describe: **1)** the substrate; **2)** cleanliness; **3)** humidity; and, **4)** competitors, are the variables that may change from one visit to another.

The individual values are divided into multiple-choice categories on the recording form. These are described in the following subsections, with the niche position occupied by each bat species summarised from the limited data held on the *BRHK Rock Face Database*. As with all the multiple-choice categories, all the options that apply should be circled.

3.5.1 Internal dimensions

The internal height is the apex of the roof of the PRF, i.e. the maximum that a bat could climb up inside the roost, and is taken from the apex of the entrance to the internal apex of the fissure or void.

The internal width is recorded from the middle of the entrance straight back on the horizontal to the back wall of the PRF. It is accepted that the width may be wider on the

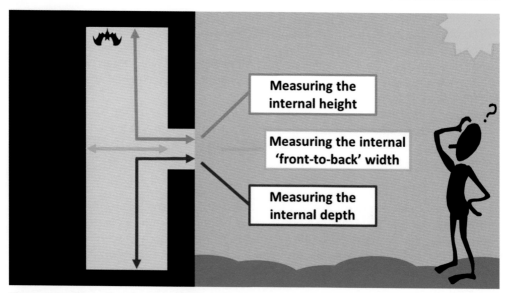

Figure 3.14 Recording the internal dimensions.

other axis (i.e. side-to-side rather than front-to-back) but the measurement is considered in respect of the bat accessing the PRF rather than occupying it once inside. In this context, the front-to-back dimension is more informative.

The internal depth is the floor of the PRF and taken from the base of the entrance. Figure 3.14 illustrates where the internal dimensions are recorded from. Table 3.9 summarises the range dimension ranges of PRF occupied by individual bat species in the limited data thus far held on the *BRHK Rock Face Database*.

Table 3.9 The dimension ranges of PRF occupied by individual bat species in the limited data thus far held on the *BRHK Rock Face Database* and photographic records.

SPECIES	INTERNAL HEIGHT RANGE (cm)	INTERNAL WIDTH RANGE (cm)	INTERNAL DEPTH RANGE (cm)
Barbastelle *Barbastella barbastellus*	0	10*	0
Serotine *Eptesicus serotinus*	0–80	29.5–80	0–6
Daubenton's bat *Myotis daubentonii*	0	29	0
Whiskered bat *Myotis mystacinus*	0–30	0–50	0
Natterer's bat *Myotis nattereri*	15–23	0–30	0
Common pipistrelle *Pipistrellus pipistrellus*	0–120	0–60	0
Soprano pipistrelle *Pipistrellus pygmaeus*	0	20	0
Brown long-eared bat *Plecotus auritus*	0–40	18–100	0

*Value estimated from photographic record.

3.5.2 Apex shape

The apex shapes in rock face situations are generally: **a)** spire-like and conical/pyramidal; **b)** peak or wedges like a pitched roof; **c)** essentially flat; or, **d)** unknown because the endoscope does not reach far enough to see.

Thus far, due to revisions in the recording process as we learned what could and could not be achieved, there is insufficient data to search for any trends in what different rock types and PRF types tend to offer, and whether individual bat species favour one sort of apex to another.

3.5.3 Internal substrate

Tree-roost data suggest that over a relatively short space of time the presence of roosting bats is associated with a change in the internal substrate of the PRF, and this appears to be more pronounced with certain species (see BTHK 2018). As with the odour cues the same has yet to be perceptible with roosts in rock faces, and at this stage of the project the internal substrate values are more a matter of what individual rock types and individual PRF types typically offer.

The values on the top row are straightforward and should be easy to identify, but the clean and messy rows will benefit from further explanation, as follows: these two rows, each with their four values, are married with both rows having one primary and three secondary attributes. It may be that only the primary attribute is applicable, e.g. clean, and the secondary attributes are not present. Or it may be that the primary is associated with one or more attributes, e.g. clean, waxy and polished. It may even be that parts of the same internal area are clean and others are messy. The recorder circles all the attributes that apply.

Thus far, as with the apex shape, we had to revise the recording process as we learned what could and could not be achieved. As a result, there is insufficient data to provide a useful summary.

3.5.4 Internal humidity

Yet again, tree-roost data have demonstrated that individual bat species display tolerances to different broad humidity regimes at different times of year. The basic classification is as follows:

- » **Dry** – arid with no suggestion of moisture;
- » **Damp** – the surface darkened by moisture but with no droplets or flow of water;
- » **Wet** – obvious beads, droplets, pooled and or flow of water;
- » **Green algae** – algal growth that suggests continuous dampness.

The humidity in the roosts occupied by individual bat species is presented as a proportion in Table 3.10.

The data suggest that dry environments are favoured. However, it may be that a visual classification is missing subtleties that are not perceptible by sight alone. It may therefore be worthwhile using a therma-hygrometer with a probe that can be inserted into the crevice to record the internal humidity (and temperature too).

3.5.5 Competitor species

The competitors encompass any other fauna that might be happy to share the PRF with bats, or find themselves evicted by what they perceive as unwelcome squatters.

Some invertebrate and bird species have been found useful for informing rapid suitability assessments of PRF in trees (see BTHK 2018). However, although in south-western caves there is an easy litmus test (the cave spider *Meta menardi* – if it is in a cave, then the

Table 3.10 The humidity inside PRF occupied by individual bat species in the limited data thus far held on the *BRHK Rock Face Database*.

SPECIES	DRY	DAMP	WET	GREEN ALGAE
Barbastelle *Barbastella barbastellus*	✓ – 100% (n=1)	—	—	—
Serotine *Eptesicus serotinus*	✓ – 92% (n=11)	✓ – 8% (n=1)	—	—
Daubenton's bat *Myotis daubentonii*	✓ – 50% (n=1)*	✓ – 50% (n=1)	—	—
Whiskered bat *Myotis mystacinus*	✓ – 67% (n=2)	✓ – 33% (n=1)	—	—
Natterer's bat *Myotis nattereri*	✓ – 80% (n=4)	✓ – 20% (n=1)	—	—
Common pipistrelle *Pipistrellus pipistrellus*	✓ – 100% (n=40)	—	—	—
Soprano pipistrelle *Pipistrellus pygmaeus*	✓ – 100% (n=2)	—	—	—
Brown long-eared bat *Plecotus auritus*	✓ – 95% (n=18)	✓ – 5% (n=1)	—	—

* Deduced from photographic record.

cave is suitable for lesser horseshoe bats *Rhinolophus hipposideros*), at the moment the data are insufficient to determine any useful associations that might be used as surrogates to inform a rapid suitability assessment. In fact, an initial review of rock faces as a habitat found that even cursory accounts were limited to nesting and roosting birds; there were no other useful accounts for any other group. A summary list of the birds that nest and roost inside rock faces (i.e. they look for cracks, breaks and off-widths within which they build nests and take shelter) is provided in Table 3.11.

More monthly replicates are desperately needed and not just the usual positive bat roost data: we need to know what is in the PRF when the bats are absent.

> **Note:** Jackdaws can form large and noisy nesting colonies in rock faces. They will occupy situations up to 70 m from the ground, and their nests may be up to 6 m from the cavity entrance (Folk 1968; Antikainen 1978). In quarries we often encounter them at *c.*10 m from the ground, which means they can occupy a wide range of the face, and be in a long way back from the surface.
>
> This species has been reported attacking noctule and long-eared bats at dusk near tower blocks in Slovakia (Mikula 2013). In half the attacks the birds were successful in bringing the bats down. It will be interesting to see if the presence of jackdaws precludes the presence of bats in the same rock face.

3.5.6 Recording whether a comprehensive inspection was possible

This is a dichotomy. For the answer to be yes, the 1st Recorder has to be confident that every bit of the internal area has been searched. In trees this is easy, in rock faces it is much more common for the 1st Recorder to finish uncertain.

Notwithstanding, to date all the roosts in rock faces that are recorded on the *BRHK Rock Face Database*, all six photographic records, and all other reports have described PRF

Table 3.11 The bird species that are known to roost inside rock faces in either or both coastal or inland situations, and the period in which they are usually present, compiled using Reade & Hosking (1974); Sharrock (1976); Mason & Lyczynski (1980); McKay (1996); and, Holyoak (2009).

SPECIES	COASTAL FACES	INLAND FACES	IN OCCUPANCY
Puffin *Fratercula arctica*	✓	X	Apr–Aug
Rock dove *Columba livia*	✓	✓	Jan–Dec
Stock dove *Columba oenas*	✓	✓	Mar–Sep
Barn owl *Tyto alba*	✓	✓	Jan–Dec
Little owl *Athene noctua*	✓	✓	Apr–Jul
Tawny owl *Strix aluco*	?	✓	Feb–Jun
Swift *Apus apus*	✓	✓	May–Aug
Swallow *Hirundo rustica*	✓	✓	May–Sep
Pied wagtail *Motacilla alba*	✓	✓	Apr–Aug
Wren *Troglodytes troglodytes*	✓	✓	Apr–Jul
Black redstart *Phoenicurus ochruros*	✓	✓	Apr–Jul
Wheatear *Oenanthe oenanthe*	?	✓	Apr–Jun
Magpie *Pica pica*	✓	?	Mar–Jun
Chough *Pyrrhocorax pyrrhocorax*	✓	✓	Apr–Jun
Jackdaw *Corvus monedula*	✓	✓	Apr–Jun
Starling *Sturnus vulgaris*	✓	✓	Apr–Jun
House sparrow *Passer domesticus*	✓	✓	Apr–Aug
Tree sparrow *Passer montanus*	?	✓	Apr–Aug

that were discrete in themselves and could be comprehensively inspected with either a torch or an endoscope with a standard 1 m long snake.

After getting endoscope extensions jammed sufficiently often for it to have caused significant delays and expense, we now have a rule that if the inspection cannot be completed with a standard 1 m snake, we simply state that a comprehensive inspection was not possible. Experience has shown that attempting to negotiate corners inside a narrow crevice with 2 m+ of extension so rarely achieves a satisfactorily controlled result that it would be irresponsible to continue: the risk of damaging the PRF or injuring something occupying it is just too great.

3.6 The rock face bat species summaries

To help people prepare for a day out by programming their semantic memory, and to analyse their results when they return, a series of species-specific summary accounts is provided with accompanying figures. These summaries are condensed to triage a search for PRF in specific rock-face habitats, and to inform a rapid assessment of PRF suitability for the specific bat species in the absence of the bats themselves.

> **Note:** The figures do not focus into the specific PRF: this is intentional. The idea is to give the reader a 'feel' for the phyz of the environment in which roosts might be present and to jog memories of other such situations that the reader might have encountered on family walks, etc. Hopefully the figures will serve to inspire and motivate the reader to go and have a look themselves.

3.6.1 Barbastelle *Barbastella barbastellus* rock face roosting account

The barbastelle rock face roosting account is constructed from one photographic record that was made in France.

Table 3.12 Barbastelle *Barbastella barbastellus* rock face roosting account.

NUMBER OF ROOSTS FROM WHICH THE SUMMARY IS CONSTRUCTED				1
SEASONS OCCUPIED	Winter: ?	Spring-flux: ?		Pregnancy: ✓
	Nursery: ?	Mating: ?		Autumn-flux: ?
	 The number of records of barbastelle roosts in rock faces occupied in each season of the year			
LANDFORMS OCCUPIED	Escarpment: ?	Crag: ✓		Tor: ?
	Gorge: ?	Boulder/block: ?	Cutting: ?	Quarry: ?
ROCK TYPE	Schist			
ASPECTS OF FACES HOLDING ROOSTS	W			
PRF TYPE EXPLOITED AS ROOST SITES	Crack: ?	Break: ✓	Flake: ?	Mono: ?
	Bidoigt: ?	Pocket: ?	Bucket: ?	Off-width: ?
HEIGHT RANGE OF ROOSTS	1.5 m (estimated from photographic record)			

ROOST ENTRANCE	Height range	2 cm		
	Width range	20 cm		
	Angle range	150 degrees		
BATS	Roost position in relation to the roost entrance	Above: 0%	In front/to the side: 100%	Below: 0%
	Maximum distance from entrance	Above: N/A	In front/to the side: 5 cm	Below: N/A
PRF INTERNAL	Height range	0 cm		
	Width range	10 cm		
	Depth range	0 cm		
HUMIDITY INSIDE ROOSTS	Dry	Damp	Wet	Green algae
	✓ – 100%	—	—	—

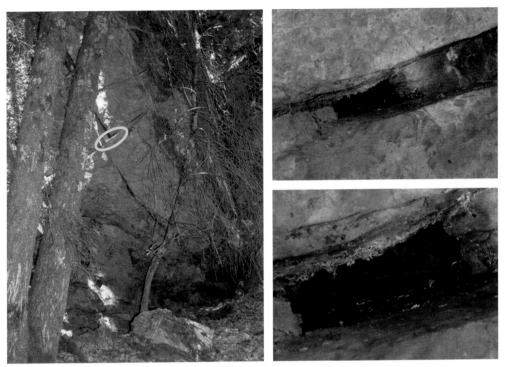

Figure 3.15 A barbastelle *Barbastella barbastellus* roost in a break. (All images: J. Matthews)

3.6.2 Serotine *Eptesicus serotinus* rock face roosting account

The serotine rock face roosting account is constructed from two detailed records and two photographic records; there were multiple records from one roost.

Table 3.13 Serotine *Eptesicus serotinus* rock face roosting account.

NUMBER OF ROOSTS FROM WHICH THE SUMMARY IS CONSTRUCTED			4	
SEASONS OCCUPIED	Winter: ✓	Spring-flux: ✓		Pregnancy: ✓
	Nursery: ✓	Mating: ?		Autumn-flux: ?

The number of records of serotine roosts in rock faces occupied in each season of the year

LANDFORMS OCCUPIED	Escarpment: ✓	Crag: ?		Tor: ?
	Gorge: ✓	Boulder/block: ?	Cutting: ?	Quarry: ✓
ROCK TYPE	Limestone and sandstone			
ASPECTS OF FACES HOLDING ROOSTS	N, S and E			
PRF TYPE EXPLOITED AS ROOST SITES	Crack: ✓	Break: ?	Flake: ✓	Mono: ?
	Bidoigt: ✓	Pocket: ?	Bucket: ?	Off-width: ?
HEIGHT RANGE OF ROOSTS	Overall range 1.45–20 m (estimated from photographic record)			

ROOST ENTRANCE	Height range	5–33 cm		
	Width range	2–6 cm		
	Angle range	90–100 degrees		
BATS	Roost position in relation to the roost entrance	Above: 0%	In front/to the side: 100%	Below: 0%
	Maximum distance from entrance	Above: N/A	In front/to the side: 25 cm	Below: N/A
PRF INTERNAL	Height range	0–80 cm		
	Width range	29.5–80 cm		
	Depth range	0–6 cm		
HUMIDITY INSIDE ROOSTS	Dry	Damp	Wet	Green algae
	✓ – 92%	✓ – 8%	—	—

Figure 3.16   A serotine *Eptesicus serotinus* roost in a crack. (Both images: H. Starkie)

Figure 3.17   A serotine *Eptesicus serotinus* roost in a bidoigt. (All images: H. Starkie)

Figure 3.18 A serotine *Eptesicus serotinus* roost behind a flake. (All images: S. Dyer)

3.6.3 Daubenton's bat *Myotis daubentonii* rock face roosting account

The Daubenton's bat rock face roosting account is constructed from one detailed record and one photographic record. Unfortunately, only the roost in the break can be illustrated.

Table 3.14 Daubenton's bat *Myotis daubentonii* rock face roosting account.

NUMBER OF ROOSTS FROM WHICH THE SUMMARY IS CONSTRUCTED				2
	Winter: ✓		Spring-flux: ?	Pregnancy: ?
	Nursery: ✓		Mating: ?	Autumn-flux: ?
SEASONS OCCUPIED	The number of records of Daubenton's bat roosts in rock faces occupied in each season of the year			
LANDFORMS OCCUPIED	Escarpment: ?		Crag: ?	Tor: ?
	Gorge: ✓	Boulder/block: ?	Cutting: ?	Quarry: ?
ROCK TYPE	Gneiss and sandstone			
ASPECTS OF FACES HOLDING ROOSTS	SE and NW			
PRF TYPE EXPLOITED AS ROOST SITES	Crack: ✓	Break: ✓	Flake: ?	Mono: ?
	Bidoigt: ?	Pocket: ?	Bucket: ?	Off-width: ?
HEIGHT RANGE OF ROOSTS	2.2–20 m (estimated from photographic record)			
ROOST ENTRANCE	Height range	2–> 50 cm (estimated from photographic record)		
	Width range	2–63 cm		
	Angle range	10–90°		
BATS	Roost position in relation to the roost entrance	Above: 0%	In front/to the side: 100%	Below: 0%
	Maximum distance from entrance	Above: N/A	In front/to the side: 13 cm	Below: N/A
PRF INTERNAL	Height range	0 cm		
	Width range	0 cm		
	Depth range	29 cm		
HUMIDITY INSIDE ROOSTS	Dry	Damp	Wet	Green algae
	✓ – 50%	✓ – 50%	—	—

Figure 3.19 A Daubenton's bat *Myotis daubentonii* roost in a break. (All images: H. Starkie)

3.6.4 Whiskered bat *Myotis mystacinus* rock face roosting account

The whiskered bat rock face roosting account is constructed from three detailed records.

Table 3.15 Whiskered bat *Myotis mystacinus* rock face roosting account.

NUMBER OF ROOSTS FROM WHICH THE SUMMARY IS CONSTRUCTED			3	
SEASONS OCCUPIED	Winter: ✓	Spring-flux: ?	Pregnancy: ✓	
	Nursery: ?	Mating: ?	Autumn-flux: ?	

The number of records of whiskered bat roosts in rock faces occupied in each season of the year

LANDFORMS OCCUPIED	Escarpment: ?		Crag: ?		Tor: ?
	Gorge: ✓	Boulder/block: ?	Cutting: ?		Quarry: ✓
ROCK TYPE	Sandstone				
ASPECTS OF FACES HOLDING ROOSTS	SW and NW				
PRF TYPE EXPLOITED AS ROOST SITES	Crack: ✓	Break: ✓	Flake: ✓		Mono: ?
	Bidoigt: ?	Pocket: ?	Bucket: ?		Off-width: ?
HEIGHT RANGE OF ROOSTS	1.48–5 m				

ROOST ENTRANCE	Height range	1.5–32 cm
	Width range	4–50 cm
	Angle range	10–90°

BATS	Roost position in relation to the roost entrance	Above: 67%	In front/to the side: 33%	Below: 0%
	Maximum distance from entrance	Above: 20 cm	In front/to the side: 6 cm	Below: N/A

PRF INTERNAL	Height range	0–30 cm
	Width range	0–50 cm
	Depth range	0 cm

HUMIDITY INSIDE ROOSTS	Dry	Damp	Wet	Green algae
	✓ – 67%	✓ – 33%	—	—

Figure 3.20 A whiskered bat *Myotis mystacinus* roost in a crack. (Both images: R. Bell)

Figure 3.21 A whiskered bat *Myotis mystacinus* roost behind a flake. (Both images: R. Bell)

3.6.5 Natterer's bat *Myotis nattereri* rock face roosting account

The Natterer's bat rock face roosting account is constructed from five detailed records and one photographic record.

Table 3.16 Natterer's bat *Myotis nattereri* rock face roosting account.

NUMBER OF ROOSTS FROM WHICH THE SUMMARY IS CONSTRUCTED			6

SEASONS OCCUPIED	Winter: ✓	Spring-flux: ?	Pregnancy: ✓
	Nursery: ?	Mating: ?	Autumn-flux: ✓

The number of records of Natterer's bat roosts in rock faces occupied in each season of the year

LANDFORMS OCCUPIED	Escarpment: ?	Crag: ✓		Tor: ?
	Gorge: ✓	Boulder/block: ?	Cutting: ✓	Quarry: ✓

ROCK TYPE	Sandstone and schist			
ASPECTS OF FACES HOLDING ROOSTS	S and SW			

PRF TYPE EXPLOITED AS ROOST SITES	Crack: ✓	Break: ✓	Flake: ✓	Mono: ?
	Bidoigt: ?	Pocket: ?	Bucket: ?	Off-width: ✓

HEIGHT RANGE OF ROOSTS	Overall range 0.8–6 m		

ROOST ENTRANCE	Height range	2–80 cm	
	Width range	2–25 cm	
	Angle range	20–110°	

BATS	Roost position in relation to the roost entrance	Above: 50%	In front/to the side: 50%	Below: 0%
	Maximum distance from entrance	Above: 20 cm	In front/to the side: 10 cm	Below: N/A

PRF INTERNAL	Height range	15–23 cm	
	Width range	0–30 cm	
	Depth range	0 cm	

HUMIDITY INSIDE ROOSTS	Dry	Damp	Wet	Green algae
	✓ – 80%	✓ – 20%	—	—

Figure 3.22 A Natterer's bat *Myotis nattereri* roost in a crack. (All images: R. Bell)

Figure 3.23 A Natterer's bat *Myotis nattereri* roost behind a flake. (Both images: R. Bell)

3.6.6 Common pipistrelle *Pipistrellus pipistrellus* rock face roosting account

The common pipistrelle rock face roosting account is constructed from 23 detailed records; there were multiple records from four roosts.

Table 3.17 Common pipistrelle *Pipistrellus pipistrellus* rock face roosting account.

NUMBER OF ROOSTS FROM WHICH THE SUMMARY IS CONSTRUCTED			23		
SEASONS OCCUPIED	Winter: ✓		Spring-flux: ✓		Pregnancy: ✓
	Nursery: ✓		Mating: ✓		Autumn-flux: ✓
	The number of records of common pipistrelle roosts in rock faces occupied in each season of the year				
LANDFORMS OCCUPIED	Escarpment: ✓		Crag: ✓		Tor: ?
	Gorge: ?	Boulder/block: ?		Cutting: ?	Quarry: ✓
ROCK TYPE	Gritstone, limestone and sandstone				
ASPECTS OF FACES HOLDING ROOSTS	SE, S, SW, W and NW				
PRF TYPE EXPLOITED AS ROOST SITES	Crack: ✓	Break: ✓		Flake: ✓	Mono: ?
	Bidoigt: ?	Pocket: ?		Bucket: ?	Off-width: ?
HEIGHT RANGE OF ROOSTS	Overall range 1–9 m with over 50% of roosts in the range 1.45–4.5 m				
ROOST ENTRANCE	Height range	1.5–1,500 cm			
	Width range	1–250 cm			
	Angle range	10–170°			
BATS	Roost position in relation to the roost entrance	Above: 18%	In front/to the side: 82%		Below: 0%
	Maximum distance from entrance	Above: 20 cm	In front/to the side: 32 cm		Below: N/A
PRF INTERNAL	Height range	0–120 cm			
	Width range	0–60 cm			
	Depth range	0 cm			
HUMIDITY INSIDE ROOSTS		Dry	Damp	Wet	Green algae
		✓ – 100%	—	—	—

Figure 3.24 A common pipistrelle *Pipistrellus pipistrellus* roost in a crack. (All images: Ian Wright)

Figure 3.25 A common pipistrelle *Pipistrellus pipistrellus* roost in a break. (All images: R. Bell)

Figure 3.26   A common pipistrelle *Pipistrellus pipistrellus* roost behind a flake. (Both images: R. Bell)

3.6.7 Soprano pipistrelle *Pipistrellus pygmaeus* rock face roosting account

The soprano pipistrelle rock face roosting account is constructed from one detailed record; there were two records from this roost.

Table 3.18 Soprano pipistrelle *Pipistrellus pygmaeus* rock face roosting account.

NUMBER OF ROOSTS FROM WHICH THE SUMMARY IS CONSTRUCTED			1
SEASONS OCCUPIED	Winter: ✓	Spring-flux: ?	Pregnancy: ?
	Nursery: ?	Mating: ?	Autumn-flux: ✓

The number of records of soprano pipistrelle roosts in rock faces occupied in each season of the year

LANDFORMS OCCUPIED	Escarpment: ?	Crag: ?		Tor: ?
	Gorge: ?	Boulder/block: ?	Cutting: ✓	Quarry: ?
ROCK TYPE	Sandstone			
ASPECTS OF FACES HOLDING ROOSTS	SW			
PRF TYPE EXPLOITED AS ROOST SITES	Crack: ✓	Break: ?	Flake: ?	Mono: ?
	Bidoigt: ?	Pocket: ?	Bucket: ?	Off-width: ?
HEIGHT RANGE OF ROOSTS	3 m			
ROOST ENTRANCE	Height range	15 cm		
	Width range	2 cm		
	Angle range	110°		
BATS	Roost position in relation to the roost entrance	Above: 0%	In front/to the side: 100%	Below: 0%
	Maximum distance from entrance	Above: N/A	In front/to the side: 10 cm	Below: N/A
PRF INTERNAL	Height range	0 cm		
	Width range	20 cm		
	Depth range	0 cm		
HUMIDITY INSIDE ROOSTS	Dry	Damp	Wet	Green algae
	✓ – 100%	—	—	—

Figure 3.27 A soprano pipistrelle *Pipistrellus pygmaeus* roost in a crack. (Both images: R. Bell)

3.6.8 Brown long-eared bat *Plecotus auritus* rock face roosting account

The brown long-eared bat rock face roosting account is constructed from 11 detailed records and one photographic record; there were multiple records from one roost.

Table 3.19 Brown long-eared bat *Plecotus auritus* rock face roosting account.

NUMBER OF ROOSTS FROM WHICH THE SUMMARY IS CONSTRUCTED				12	
SEASONS OCCUPIED	Winter: ✓		Spring-flux: ✓		Pregnancy: ✓
	Nursery: ?		Mating: ?		Autumn-flux: ✓
	The number of records of brown long-eared bat roosts in rock faces occupied in each season of the year				
LANDFORMS OCCUPIED	Escarpment: ?		Crag: ✓		Tor: ?
	Gorge: ✓	Boulder/block: ?		Cutting: ✓	Quarry: ✓
ROCK TYPE	Sandstone and tuff				
ASPECTS OF FACES HOLDING ROOSTS	N, NE, E and SW				
PRF TYPE EXPLOITED AS ROOST SITES	Crack: ✓	Break: ?		Flake: ✓	Mono: ?
	Bidoigt: ✓	Pocket: ?		Bucket: ?	Off-width: ?
HEIGHT RANGE OF ROOSTS	0.75–5 m with over 50% of roosts in the range 1.08–3.9 m				
ROOST ENTRANCE	Height range	3–300 cm			
	Width range	1.5–28 cm			
	Angle range	0–120°			
BATS	Roost position in relation to the roost entrance	Above: 25%		In front/to the side: 75%	Below: 0%
	Maximum distance from entrance	Above: 25 cm		In front/to the side: 35 cm	Below: N/A
PRF INTERNAL	Height range	0–45 cm			
	Width range	18–100 cm			
	Depth range	0 cm			
HUMIDITY INSIDE ROOSTS	Dry	Damp		Wet	Green algae
	✓ – 95%	✓ – 5%		—	—

Figure 3.28 A brown long-eared bat *Plecotus auritus* roost in a crack. (All images: H. Starkie)

Figure 3.29 A brown long-eared bat *Plecotus auritus* roost behind a flake. (All images: H. Starkie)

3.7 The Point

If you did not find a bat but you have the form you can compare it to the information in this chapter to get an idea of how likely it is that the PRF is suitable.

Furthermore, if you complete the recording form, you will have a priceless piece of data whether there is a bat in the PRF that day or not. If you submit it to the *BRHK Rock Face Database* you will not only advance scientific knowledge, you will also have a public record of your experience which you can cite when you are going after a pay rise from your employer or at a job interview when you decide not to continue working for a corporate biodiversity pimp who does not reward scientific endeavour and does not value data over money.

Continue to record and you will begin to play the game, and being able to play the game makes your work more fun.[15]

CHAPTER 4

Loose rock: characterising and recording the landforms and the Potential Roost Features they may hold

Henry Andrews & James McGill

In this chapter	
Preamble	Summary recap of Chapters 2 and 3; Introducing scree; Introducing talus; Introducing blockfield; Introducing the clasts/rocks
What	Introducing the loose rock meso-environment values; Introducing the loose rock roost account
What 1: recording the loose rock meso-environment values	The rock type; The landform; Whether the landform is natural or artificial; The landform shape; The landform topography; The landform aspect; The angle of the rock face slope; The contiguousness of the landform; The habitat on and around the landform; The presence of islands of vegetation; The proportion of clasts with moss cover; The proportions of clasts with lichen cover; Bat entry points
What 2: recording the bats	The bats
The loose rock bat species summary	Natterer's bat *Myotis nattereri* loose rock roosting account
The point	Finishing your career as a player and not a disillusioned hack
Just for interest	The multitudinous myriad of competitors that occupy loose rock landforms – Food for bats? Food for thought …

4.1 Preamble

In Chapter 2 the different rocks the reader might encounter in their part of the British Isles were identified. The landforms made of those rocks were then described so that they might be identified in the field.

In Chapter 3 the landforms that offer surface rock faces were described in the context of bat roost habitat. This approach is continued in Chapter 4 which describes the loose rock landforms that bats might exploit as a roost habitat, these comprise: **1)** scree; **2)** talus; and, **3)** blockfield.

To set the scene, the three landforms are illustrated in Photos 4.1–4.3.

Photo 4.1   A natural scree. (© ImageBROKER.com/Shutterstock)

Photo 4.2   A natural talus.

Photo 4.3 A natural blockfield. (© Circumnavigation/Shutterstock)

4.1.1 Introducing scree

A scree is an accumulation of rock fragments piled against the foot of a rock face or flowing out of a gulley. Screes form as the result of physical and chemical weathering and erosion of the face above and behind the scree slope. The individual rocks that make up a scree are known as 'clasts' and these range from pebbles, through cobbles and up to blocks and boulders.

Screes are sloped landforms. Depending upon the rock type and the situation the clasts may hold together against the face to leave the scree with a convex surface in a broad cone shape or spread out from the face in a concave fan (see Photo 4.4). In more reclined topography the faces are more-or-less smooth and flat, as in Photo 4.5. Finally, some other screes appear subtly stepped in a series of shelves, as in Figure 4.1.

The stepped screes encountered during the project investigations were small and often partially vegetated. It is unknown whether the stepped effect is due to climatic processes, or the vegetation, or animals walking across the face.

Any mountaineers reading in southern Europe may use the word talus instead of the word scree, and indeed the two words have been used synonymously: the entry for talus in the Penguin Dictionary of Geology says 'see scree'.

A search of Wikipedia informs us that scree is the Old Norse for 'landslide' and talus is French for 'embankment'. The two words are used by geologists and mountaineers to mean the same thing: a slope of loose rock that has detached from a rock face as a result of freeze–thaw weathering (i.e. congelifraction) and rolled downhill to leave a fan or cone-shaped exposure on a mountainside at the base of a crag or escarpment.

Notwithstanding, anyone who has looked at screes in a wide range of situations will very soon realise that they do not all fit the 'textbook' image previously provided in

Photo 4.4 Cones and fans. (© Merry Jane/Shutterstock)

Photo 4.5 A smooth scree.

Figure 4.1 A subtly stepped scree.

Photo 4.1. In the textbook cases there is a convex cone or a scree flow that is segmented, with a generally straight upper slope and a lower basal concavity, the upper being a transport surface and the lower being the gradual settlement of the deposited material (Francou & Manté 1990).

These are examples of a gradual process that has taken place over millennia, but what of sudden rockfalls resulting from catastrophic collapse? The rocks at the base of such a collapse are dominated by much larger boulders. These features look nothing like screes and from a bat's perspective will offer a very different environment: rather than the superabundance of shallow crevices, these huge boulder piles offer a network of deep fissures and small to moderate-sized voids.

In order to separate these environments, we need another word and in the context of this project: so, we are commandeering talus to conform with two useful descriptions given by Parry (2011) and Wilson (2012).

4.1.2 Introducing talus

Parry (2011) describes talus slopes as interlocking angular blocks of over 2 m³ on steep slopes with a face angle of over 30°. His description suggests the features are generally restricted to within 30 m of the foot of a rock face but may be further away if the boulders have bounced. Wilson (2012) describes the landform as having spaces between that may be large enough for a person to enter.

Scree exhibits a distinct fall sorting: the clasts flow away from the face smallest to largest, with the boulders at the base of the fan. But not all loose rock landforms accord with that structure. This is particularly evident where there has been a massive collapse in a lowland situation; here the largest boulders are at the base and the smaller ones spread out like detritus thrown out from a meteor strike.

For the purpose of the BRHK project, the word talus will be applied to landforms that are the result of catastrophic collapse. These landforms may be encountered in isolation, or in association with scree, and the distinction will allow even combination landforms to be divided and any bat roost present to be described in meaningful context. An example of this is the scree that has been overtaken by a landslide resulting in two talus patches shown in Figure 4.2.

In some talus the surface is just like a scree on steroids, as in Photo 4.6. In other situations, the boulders are sufficiently large and slab-like to create small caves. Talus caves are interconnected spaces between rock that are large enough for a person to enter and move into a position beyond daylight. In Sweden they are common, and some are thousands of metres in length (Halliday 2004). It is not impossible that there are talus caves extending tens of metres exist in the UK which have escaped the interest of cavers but not roosting bats. Talus caves should be subject to a risk assessment before a surveyor considers entering and appropriate PPE should be worn. The rocks are not fixed to each other and attempting to wriggle in might easily dislodge something. Talus cavelets are illustrated in Figure 4.3.

And having defined the intermediate landform, we can look at the last of the three: blockfields.

Figure 4.2 A combination of scree and talus. (Original image: © Evgeny Haritonov/Shutterstock)

Photo 4.6 The block-strewn surface of a talus.

Figure 4.3 Talus cavelets.

Photo 4.7 The rounded boulders of a weathered blockfield. (© Ion Sebastian/Shutterstock)

4.1.3 Introducing blockfield

An excellent definition of a blockfield is achieved by simply translating the German name for the feature: *felsenmeer* (sea of rocks).

Blockfields are the result of the freezing of a surface layer of rock to leave the substrate cracked and broken. Initially the material comprises jagged and angular blocks, but (depending on the rock type) the blocks weather or are eroded by glacial movement and meltwater flow, to leave rounded boulders, as in Photo 4.7.

Unlike scree and talus which are sloped with the rocks typically resting against a face, blockfields are level and only form on slopes that have an angle less than 25°.

4.1.4 Introducing the clasts/rocks

While it is not impossible that bats may burrow into a surface, and Schober & Grimmberger (1993, 1997) even provide a photograph of a Daubenton's bat *Myotis daubentonii* that appears to have burrowed into soil, it is not something that they appear particularly well designed to do. Working to the premise that bats will generally prefer to exploit a pre-existing gap, the size and shape of the rocks that make up a loose rock landform decide whether the surface has openings into which a bat might crawl in order to find a dark place to hide.

Geologists describe three rock sizes/shapes, from smallest to largest: **1)** pebbles; **2)** cobbles; **3)** boulders; and, **4)** blocks. However, in an attempt to define a set of recording values that were meaningful in describing what was seen in the field in terms of bat accessibility, these four sizes/shapes had to be augmented with two more: **1)** chips; and,

2) plates, which are illustrated with an asterisk in the list of the six clast sizes and shapes we will use to characterise screes below:

» **Pebbles** – rounded rock fragments between 2 mm and 64 mm in diameter (Kearey 1996);

» **Chips*** – broadly flat flakes of rock that are between 2 mm and 150 mm in diameter on any one plane (BRHK concept);

» **Cobbles** – rounded rock fragments between 64 mm and 256 mm in diameter (Kearey 1996);

» **Plates*** – broadly flat plates of rock that are over 150 mm in diameter (BRHK concept);

» **Boulders** – rounded rock fragments above 256 mm in diameter (Kearey 1996);

» **Blocks** – angular rock fragments above 256 mm in diameter (Kearey 1996).

In terms of scale, boulders and blocks are easy to imagine but the difference in the range of pebbles, chips and cobbles and pates is less so. Figure 4.4 shows the size ranges using an endoscope for scale.

All the definitions having been set out, we may now move on to the 'what' of a biological record in the context of loose rock landforms.

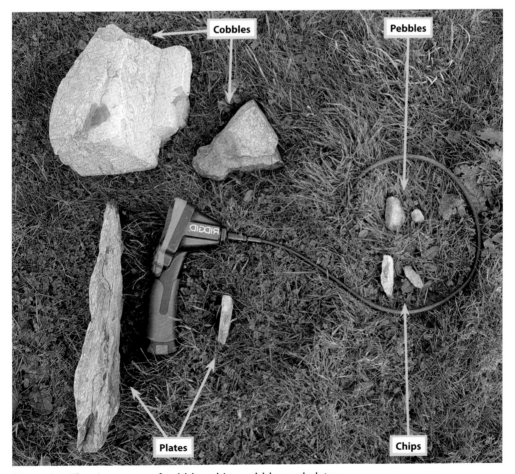

Figure 4.4 The size ranges of pebbles, chips, cobbles and plates.

4.2 What

To recap Chapter 1, the 'what' is all the meso- and micro-scale environmental values.

Continuing with the rationale set out in Chapters 1 and 3, the framework for the data review and the chapter layout will follow the order of the standardised *BRHK Loose Rock Database* recording form. This means that the information collated and presented will have a practical application in the field.

To recap: in Chapter 1 we looked at the 'who', 'where' and 'when' of a biological record. Chapter 3 comprised 'what' else we needed to record for a useful record of a specific surface rock face, and this chapter will do the same for loose rock landforms. The scope of this chapter is narrower than that of Chapter 3 because the data that has been gathered thus far has been basic, and this continues to be true with what we can usefully record in the field.

The level of detail we can record will be constrained by the complexity of the three landforms, and the fragility of screes.

Attempting to map and record every possible situation in which a bat might roost within loose rock is impractical. In addition, the unstable nature of scree is such that where a surveyor attempts to search one there is a risk of injury and even death to any bat occupying it, both from the weight of the surveyor on the loose rocks and also from the collapse of the loose structure as the surveyor attempted to move individual components. The risk of injury to the surveyor is also a consideration, from slips and falls, through localised landslips and even from larger rockfalls. This is pertinent to scree but also talus: the potential for bone-breaking injuries is significant when attempting to clamber between boulders, and there is the added risk of crushing.

In the context of these landforms physical inspections may therefore be unwise, and potentially dangerous.

Yet the three loose rock landforms may be encountered from mountain slopes down into the base of quarries and, despite their wide geographical range, we find that satisfying Project Objective 2 – 'The collation of published descriptions of bat roosting ecology on, among and in natural and semi-natural rock habitats' – is a light task because loose rock habitats are the no man's land of bat roosting ecology: we really have very little idea just how much these features are used in the British Isles.

What we can say with certainty is that loose rock is used as roost habitat by at least four bat species that occur in the British Isles, and if we can identify the broad physical and environmental characteristics of the situations in which loose rock is used, we can replicate them. And that replication may be a relatively cheap and easy way to create roost habitat.

At present there is insufficient evidence to set out the values that might be recorded to assess whether or not loose rock is suitable for a particular bats species to exploit as a roost, or whether it is in fact a roost already. However, the chapter collates what little information there is, and advances a framework of values that might be recorded by anyone who finds a bat roost in loose rock. The idea is that: **a)** new encounters are recorded in a standardised format using values that are meaningful in the context of a visual assessment of a specific landform; and, **b)** individual landforms are recorded so that potentially important sites get the attention they deserve.

4.2.1 The recording form

To refresh your memory, the *BRHK Loose Rock Database* recording form is provided again in Table 4.1.

The values comprise a basic landform record, and encompass the meso-environment variables that characterise the situation and form of the habitat, and of course the situation any bat(s) might be roosting in on the day. This can be considered the 'mapping record' too, because with an annotated image it would help someone else to find to the landform and assess whether it had physically changed.

Due to the risks to bats and surveyors there is no 'advanced record' and it may be unwise to attempt any intrusive investigations, particularly if such an investigation would be likely to constitute a destructive search.[16]

Table 4.1 The *BRHK Loose Rock Database* recording form.

WHO	1st Recorder:				2nd Recorder:		
WHERE	**Site name:** see Chapter 1						
	Grid reference: see Chapter 1						
WHEN	**Date:** see Chapter 1						
WHAT	Rock type (e.g. granite):						
	Landform type: scree / talus / blockfield						
	Landform composition – Proportion of surface occupied by:	Pebbles:	Chips:	Cobbles:	Plates:	Boulders:	Blocks:
	Landform origin: natural / artificial						
	Landform shape: cone / fan / other						
	Landform topography: flat / stepped / convex / concave						
	Landform aspect (e.g. North or N/A if it is simply a pile of rock):						
	Angle of slope:						
	Contiguousness: continuous / discrete within network / discrete and isolated						
	Habitat around and on the landform (phase 1):						
	% island vegetation present (i.e. is anything growing within the landform):						
	% moss cover:						
	% of clasts* without moss cover that hold lichens:						
	Entry points:	Obvious gaps into which a bat might climb			No obvious spaces; bat would have to burrow into substrate		
ROOST ACCOUNT		Bat species:					
		Minimum number of bats:					
		Maximum distance from edge of landform to roost position (m):					
		Awake or torpid:					
		Roost depth (cm):**					
		Describe any visual field-signs that might have given the presence of a roost away just by looking at the landform:					

* A clast is a particle of rock derived by weathering and erosion (Kearey 1996).
** Only if being measured as part of specific academic study or justified by circumstances.

4.2.2 Introducing the loose rock meso-environment values

The meso-environment is what a bat encounters as it flies through the landscape and encompasses the rock type the landform is made of, the landform itself, and the situation in which it might roost.

In the context of the BRHK project the loose rock meso-environment values comprise: **1)** the rock that the landform is made of; **2)** the landform type; **3)** the proportion of the landform surface occupied by pebbles, chips, cobbles, plates, boulders and blocks; **4)** whether the landform is natural or artificial; **5)** the landform shape; **6)** the surface topography; **7)** the landform aspect (i.e. which way it faces); **8)** the angle of the slope; **9)** whether the landform is contiguous with more of the same landform in an indivisible blanket, or it is a discrete patch; **10)** the habitat around and on the landform; **11)** the proportion of vegetation growing within the landform; **12)** the proportion of the landform that is clad with moss; **13)** the proportion of the clasts/rocks without moss cover that hold lichens; and, **14)** whether there are gaps into which a bat might climb or whether the bat would have to burrow, nudging rocks aside to penetrate the substrate.

With all that recorded attention can be turned to the bat(s).

4.2.3 Introducing the loose rock roost account

Historically, accounts of bats in loose rock in surface situations have been the result of radio-tracking. And while there are accounts of bats being physically uncovered among loose rock in subterranean situations, these appear to have been the result more of accident than of structured searches.

Radio-tracking comprises the capture of free-flying bats with the subsequent attachment of a radio-transmitter to the bat, which enables the bat to be located by triangulation by two or more surveyors using radio-receivers. To date, although records made by radio-tracking have identified that bats are roosting among the clasts/rocks, and sometimes the broad location of the bat in terms of the quadrant within the overall landform, no single paper has described the exact position of the bat in terms of depth below the surface or provided a detailed description of the micro-environment the bat is occupying (although Alberdi *et al.* (2014) have come close with alpine long-eared bat *Plecotus macrobullaris*).

The loose rock recording framework is basic but should build to give a broad idea of the sorts of situation individual species are exploiting, in order to inform their value and, if desired, an attempt to replicate them.

The following sections describe the individual values so the surveyor can recognise them and record them.

Note: By grouping information taken from scientific accounts from Europe, Canada and the USA it would be possible to put together a general narrative giving a summary of how loose rock is used by bats around the globe, but this would be equivalent to 'Frankenstein's Chapter' and would require compromise at every value (as well as being irrelevant in the context of the British Isles). The decision was therefore taken to limit the accounts to bat species that occur in the British Isles. As a result, the following text refers to four sources of information, comprising:

1. Radio-tracked individual Daubenton's bats have been recorded using screes under sheltered limestone cliffs (Altringham 2003).

2. Billington, G. 2000. *Holnicote Estate – Horner Woods Bat Survey – Somerset*. Greena Ecological Consultancy, Frome

    Working with a team in Somerset, Geoff Billington recorded one radio-tagged Natterer's bat *Myotis nattereri* roosting in a sandstone scree with a natural origin surrounded by woodland, gorse *Ulex europaeus* and bracken *Pteridium aquilinum*. The scree is diamond-shaped and c.15 m high and 6 m wide. The depth of the material is unknown but it is east-facing. The roost was occupied for at least four consecutive days and several Natterer's bats were observed emerging from the scree including the tagged individual. Helpfully this report includes a photograph of the scree and an OS grid reference that pinpoints its location and that of the bat in it.

3. Michaelsen, T., Olsen, O. & Grimstad, K. 2013. Roosts used by bats in late autumn and winter at northern latitudes in Norway. *Folia Zool.* 62(4): 297–303

    This team recorded two whiskered bats *Myotis mystacinus* and a brown long-eared bat *Plecotus auritus* roosting in screes on islands off Norway. The rock the screes were composed of is not given but as far as can be deduced they are gneiss. Both are naturally formed but the habitat surrounding them is unknown, as is the landform shape, surface area, or aspect. The actual roost position was not reported nor is there any photograph that might be used to deduce any additional detail.

4. Michaelsen, T. & Grimstad, K. 2008. Rock scree – a new habitat for bats. *Nyctalus* 13: 122–126

    Working on islands off Norway, this team recorded brown long-eared bat roosting in a rock face that was covered by scree. The rock the scree was composed of is not given but as far as can be deduced it is gneiss. The scree was naturally formed but the habitat surrounding it is unknown as is the landform shape, surface area, or aspect. The actual roost position was not reported nor is there any photograph that might be used to deduce any additional detail.

N.B. It appears likely that the individual brown long-eared bat that roosting in scree that is referred to by Dietz *et al.* (2011) and Dietz & Kiefer (2016) relates to this account.

All four accounts are considered reliable records in the context of the chapter.

4.3 What 1: recording the loose rock meso-environment values

4.3.1 The rock type

The records available comprise two rock types: gneiss (Michaelsen & Grimstad 2008; Michaelsen *et al.* 2013) and sandstone (Billington 2000).

This value is easiest and most confidently recorded in the office by entering the OS grid reference into the mapping page of the British Geological Survey website[17] and simply clicking on the big red pin that marks the location.

As was explained in Chapter 3, the reason the rock is important is that different rocks may form different PRF and offer different environments due to factors such as grain, porosity and heat retention. These environments may be exploited in different ways in different seasons by different bat species.

By cataloguing the rocks, particular trends may be identified for use in a predictive analysis. This will allow the recorder to mentally prepare themselves to search for specific PRF types in different situations, and thereby use their limited time most effectively by triaging their effort to inspect the most profitable PRF types first.

4.3.2 The landform

To recap, in the context of the BRHK project there are three loose rock landforms, comprising: **1)** scree; **2)** talus; and, **3)** blockfield. In the context of the BRHK project, a scree, a talus and a blockfield will be separated as follows:

» **Scree** – For the purposes of the BRHK project, a scree is defined specifically as any natural or artificial loose rock substrate on sloping ground with an angle of slope typically above 25°, the surface of which is fragile and dominated by pebbles, chips, cobbles and/or plates. Although boulders and blocks may also be present they will be a subordinate component.

A scree might occur naturally at the base of a sheer rock face and above a blockfield. Artificial forms are often found piled against quarry faces.

Photo 4.8 illustrates the surface of a scree.

» **Talus** – For the purposes of the BRHK project, a talus is defined specifically as any natural or artificial loose rock substrate on sloping ground with an angle of slope typically above 25°, the surface of which is interlocked and dominated by blocks and boulders. Although cobbles, plates, pebbles and chips may be present, they will be a subordinate component.

A talus might occur naturally at the base of a sheer rock face but artificial forms are often found against quarry faces.

Photo 4.9 illustrates the surface of a talus.

» **Blockfields** – A blockfield is an accumulation of cobbles, plates, boulders and/or blocks on level or gently sloping ground (Kearey 1996) with an angle of slope that is typically level and certainly no more than 25°. For the purposes of the BRHK project, a blockfield will comprise any expanse of rocks that is over 50% boulders and/or blocks. The effect is to offer a void-type PRF web with a multitude of entrances into which a bat might crawl or even fly, and rest in a roost position inside and below.

Blockfields may be natural or artificial and of the three loose rock landforms they are the most stable (the rocks often being compacted against each other) and the one most likely to be vegetated.

Photo 4.10 illustrates the surface of a BRHK project blockfield.

Michaelsen & Grimstad (2008) provide a contextual close-up photograph of the clasts in one of the loose rock landforms within which their whiskered and brown long-eared

Photo 4.8 The surface of a scree.

Photo 4.9 The surface of a talus.

Photo 4.10 The surface of a blockfield.

bats were roosting, and it is a cobbled scree. Billington (2000) gave a detailed record of his Natterer's roost with an accompanying photograph. This informed a visit and a test of the BRHK Loose Rock recording criteria: we can confidently state that this is a scree of plates over chips and pebbles, and with conspicuous blocks also present.

In order to search for specific favourable attributes, the landform value on the recording form includes the proportion of the surface occupied by each clast size. By recording two values it will be possible to see whether any particular clast size offers more roost potential than any other, or whether a specific combination is required, and whether this differs with the rock type.

4.3.3 Whether the landform is natural or artificial

Away from rocky coastlines, natural examples of the loose rock landforms occur in upland situations at the base of crags, escarpments and gorges where the winter climate is colder, and also in lowland coastal situations. Artificial examples encountered during this project have been in quarries and around mines,[18] such as those shown in Figure 4.5.

It is unknown whether the account of Daubenton's bats given by Altringham (2003) relates to a natural landform. The Natterer's bat recorded by Billington (2000), the whiskered bats recorded by Michaelsen *et al.* (2013), and the brown long-eared bats recorded by Michaelsen & Grimstad (2008) and Michaelsen *et al.* (2013) were certainly recorded in natural landforms.

Figure 4.5 Top: artificial scree. Middle: talus. Bottom: blockfield. All comprise tipped limestone.

4.3.4 The landform shape

The landform shape is divided into three options that represent all the shapes that were encountered in producing the recording framework. These comprise:

» **Cone** – The typical reclined convex spire of a crag or escarpment scree and talus in a mountainous situation;

» **Fan** – The typical triangle shape of loose rock on a gorge-side and also a material piled against a quarry face;

» **Other** – You tell us what shape it is.

The only sufficiently detailed record is of the Natterer's bat roosting in a scree, and this is a diamond shape.

4.3.5 The landform topography

This value relates to the landform itself and has four values, as follows:

» **Convex** – The textbook mountain scree form;

» **Concave** – The typical topography of a mountainside fan in a gulley situation;

» **Flat** – This encompasses any situation where the surface is flat, and includes sloping flat surfaces as well as level ones;

» **Stepped** – Several small gorge-side screes have been encountered with gently stepped faces.

The only sufficiently detailed record is of the Natterer's bat recorded roosting in a scree by Billington (2000), and this has a flat surface.

4.3.6 The landform aspect

The aspect of a loose rock landform will determine how exposed it is to the sun's rays and therefore how cold it gets in the winter, how hot it gets in sunny weather, and also how much light hits it, when, and for how long each day. This meso-environmental gradient may therefore influence the micro-environment temperature, humidity and illuminance gradients. It will also influence when the face is sufficiently dark for bats to begin emerging. The aspect of the rock face is recorded using a compass and to the nearest eighth, e.g. NE.

The only sufficiently detailed record is of the Natterer's bat roosting in a scree by Billington (2000), and this has a north-eastern aspect.

4.3.7 The angle of the rock face slope

The angle of the slope is of interest both from the perspective of surface mobility and stability, and also hours of sunlight. It is recorded as a degree (°) value using a clinometer (most phone apps are fine) or by eye using a protractor (just download a photo of one onto your phone for reference in the field).

The only sufficiently detailed record is of the Natterer's bat roosting in a scree by Billington (2000), and this has a 35° angle of slope.

4.3.8 The contiguousness of the landform

This refers to the scope of the landform and whether it is: **a)** part of a continuous expanse of loose rock that skirts the foot of a crag, with those of other crags flowing down to meet it; **b)** a discrete individual within a wider network of patches that are divided by vegetation; or, **c)** an isolated individual. Figure 4.6 illustrates continuous, discrete and isolated loose rock landforms.

It may be easier to assess this situation using Google Earth.

Figure 4.6 Loose rock landforms. Top: continuous. Middle: discrete. Bottom: isolated. (Top image: © Matyas Rehak/Shutterstock)

The distinction here is really whether the bat is occupying a specific loose rock landform and a particular feature within it, or whether it simply sees a surface that it knows will offer PRF and takes the nearest available hole to where it lands.

The only sufficiently detailed record is of the Natterer's bat roosting in a scree by Billington (2000), and this is a discrete individual scree within a wider network of screes.

4.3.9 The habitat on and around the landform

This relates to whether or not the landform itself is vegetated and the habitat context in which the landform is situated. The habitat is recorded using the Handbook for *Phase 1 Habitat Survey – A Technique for Environmental Audit* (JNCC 2010) which can be downloaded free of charge.[19] However, to save some time at the start, the rock landforms fit into the following Phase 1 habitats:

» *Natural landforms:*
  - **Scree** – I1.2 – Rock exposure and waste / Natural / Scree;
  - **Talus** – I4.1 – Rock exposure and waste / Natural / Other exposure;
  - **Blockfield** – I1.4 – Rock exposure and waste / Natural / Other exposure.
» *Artificial situations:*
  - **Scree** – I2.2 – Rock exposure and waste / Artificial / Waste tip;
  - **Talus** – I4.1 – Rock exposure and waste / Artificial / Waste tip;
  - **Blockfield** – I2.2 – Rock exposure and waste / Artificial / Waste tip.

The only sufficiently detailed record is of the Natterer's bat roosting in a scree by Billington (2000), and this is:

» **I1.2.1** – Rock exposure and waste / Natural / Scree / Acid;
  Surrounded by a mosaic of:
  - **A1.1.1** – Broadleaved semi-natural woodland;
  - **C1.1** – Continuous bracken;
  - **J5** – Other habitat / gorse.[20]

Note: Although it is labouring the point made in Chapters 1 and 3, if we have the full 12-value alphanumeric OS grid reference, we can look at the habitat remotely using satellite images to see if there is anything common to roost situations occupied by specific bat species, and anything that separates them.

4.3.10 The presence of islands of vegetation

Island vegetation refers to any higher plant that is growing within the loose rock landform and potentially spreading across the surface from within. This specifically excludes mosses and lichens which have their own values, and encompasses: trees, shrubs, flowering plants, grasses, sedges, rushes, and ferns. Figure 4.7 illustrates island vegetation on screes.

The surveyor does not need to identify the species within the island, but simply note the proportion of the surface of the landform that is vegetated. The surveyor should view the surface from a bat's perspective: if you were flying 5 m above the ground what proportion of the landform surface would be visible to you?

The only sufficiently detailed record is of the Natterer's bat roosting in a scree by Billington (2000), and this has no island vegetation growing within it.

Figure 4.7 Island vegetation on screes.

4.3.11 The proportion of clasts with moss cover

Moss cover suggests damp conditions. This does not, however, suggest poor likelihood of a roost. Factors to keep in mind: **1)** the presence of moss demonstrates that the surface is stable; **2)** moss soaks up a lot of water and may therefore stop rain from penetrating deep under the rock surface; **3)** moss will insulate against wind; and, **4)** tree-roost data demonstrates that Natterer's bats, common pipistrelles *Pipistrellus pipistrellus* and soprano pipistrelles *Pipistrellus pygmaeus* will occupy damp roosts, and Daubenton's bats and brown long-eared bats will occupy roosts that are visibly wet inside (BTHK 2018). Photo 4.11 illustrates a moss-covered scree that is peppered with little gaps leading down into darkened spaces in which a bat might shelter.

4.3.12 The proportion of clasts with lichen cover

Lichens are slow to colonise and even slower to grow (Gilbert 2000). The presence of lichens on the clasts/rocks suggests stable conditions over decades. The greater the surface that is covered, the longer it has been stable. Referring back to Figure 4.1 and Photo 4.6, these show conspicuous lichen cover on clasts/rocks that have been a stable situation for

Photo 4.11 A moss-covered scree that is peppered with little gaps leading down into darkened spaces in which a bat might shelter.

Photo 4.12 A plate that has been lifted on end to illustrate that although the upper surface is lichen covered, the underside is clean sandstone which is conspicuous against the lichen-clad pebbles all around.

a long period (i.e. centuries old rather than decades). To really make the point, Photo 4.12 shows a plate that has been lifted on end to illustrate that although the upper surface is lichen covered, the underside is clean sandstone which is conspicuous against the lichen-clad pebbles all around.

4.3.13 Bat entry points

This chapter differs from Chapters 3 and 5 because there are no specific PRF types described. This is because the landform itself is really the PRF. The 'entry points' value is looking to see whether any one rock type or clast size/shape offers more in the way of visible entry points a bat might exploit to penetrate the surface without having to burrow in.

If you use an endoscope as part of your job this value is easy: just ask yourself whether you would even bother to get the endoscope out of the bag. If the answer is a confident 'yes' then there are obvious holes into which a bat might crawl. If the answer is a sigh and a 'maybe' then the answer is that a bat would likely have to burrow into the substrate.

This is important because the clast size and shape may be misleading in accessibility terms. For example, sandstone plates are fantastic for crevices, but sandstone pebbles and chips (and plates in some situations, such as slate) may leave an impenetrable surface like snakeskin.

Figure 4.8 illustrates loose rock with an abundance of entry points a bat might exploit to access a roost position below the surface, and another surface into which a bat would have to make its own access by burrowing.

Figure 4.8 Left: loose plates and cobbles with an abundance of entry points. Right: loose chips and pebbles with no obvious entry points.

Figure 4.9 The large gaps that are left between talus blocks.

Talus has bigger gaps that effectively leave voids up to cavelet size, and these can continue into what is effectively a very constrained system of passages. Figure 4.9 illustrates talus entrances.

4.4 What 2: recording the bats

4.4.1 The bats

Although the data with sufficient detail to be of practical value in the field comprise just one landform occupied by one bat species, an overall 12 bat species that occur in the British Isles are reported to exploit features in exposed surface rock-face habitat as roost sites, these comprise:

1. **Daubenton's bat** *Myotis daubentonii* (Altringham 2003);
2. **Whiskered bat** *Myotis mystacinus* (Michaelsen *et al.* 2013);
3. **Natterer's bat** *Myotis nattereri* (Billington 2000);
4. **Brown long-eared bat** *Plecotus auritus* (Michaelsen & Grimstad 2008; Michaelsen *et al.* 2013).

As explained in Chapter 3, this project is all about standardisation of records, providing proof that the records are accurate, and gathering information that will have a practical application in informing searches and artificial roost designs. To the latter end the values recorded when bats are present comprise: **a)** the species; **b)** the minimum number of bats; **c)** the shortest distance between the location the bat is roosting in and the edge of the landform; **d)** the roost depth; and, **e)** whether there were any visible field-signs on the rock surface that indicated the presence of a roost (this might not only be the presence of some conspicuous characteristic: it might also be the absence of something that was otherwise comprehensive, such as moss cover everywhere but on one side of one particular block which has a bare patch leading up into a crevice-type gap occupied by a bat).

Note: Subterranean scree and talus are exploited by:

1. **Barbastelles** *Barbastella barbastellus* (Dietz *et al.* 2011; Dietz & Kiefer 2016);
2. **Serotine** *Eptesicus serotinus* (Schober & Grimmberger 1993, 1997; Dietz *et al.* 2011; Dietz & Kiefer 2016);
3. **Bechstein's bats** *Myotis bechsteinii* (Dietz & Kiefer 2016);
4. **Brandt's bats** *Myotis brandti* (Dietz & Kiefer 2016);
5. **Daubenton's bats** (Schober & Grimmberger 1993, 1997 – suggest the bats may penetrate up to 60 cm deep; Stebbings 1988 – suggests up to 1 m deep; Dietz & Kiefer 2016);
6. **Natterer's bats** (Schober & Grimmberger 1993, 1997; Dietz *et al.* 2011; Dietz & Kiefer 2016);
7. **Whiskered bats** (Dietz & Kiefer 2016);
8. **Brown long-eared bats** (Dietz *et al.* 2011; Dietz & Kiefer 2016).

Although these are loose rock substrates, they are considered in Chapter 5 because they are underground in caves and mines. However, these accounts do at least demonstrate that in addition to the four species that have been recorded roosting in surface loose rock landforms another four are not afraid of landing on loose rock and climbing into it. If they will exploit it in subterranean situations, they physically can use it in surface situations.

One value in the list given in the last paragraph warrants further consideration: the roost depth.

Obviously, you are not expected to habitually attempt to expose roosting bats. This value is for: **a)** specific scientific investigations under licence (because to uncover the bat you may be damaging the roost feature); **b)** destructive searches under licence; and, **c)** accidental discoveries (everyone makes mistakes, and it is hoped that a bat that was accidentally uncovered would understand the need to maximise every opportunity for advancing scientific knowledge to benefit their species). Some advice on searching is provided in Chapter 6, but in fact, a non-intrusive investigation is often sufficient to make an estimate of the likely depth the bat might penetrate down into. This is because the size of the clasts decreases the deeper the surveyor digs; it is a curious phenomenon, but in general terms the boulders and blocks sit in layers of cobbles and plates, which sit on pebbles and chips. Luckman (2013) and Rixhon & Demoulin (2013) examined loose rock landforms in detail and found that even where the surface was coarse and open this was a veneer. In fact, the open surface was a sieve thought which smaller clasts gradually worked down to fill every gap.

A simplified explanation of how this comes about was pieced together from a multitude of sources and is provided below. The individual sources are not cited in the following paragraph as to do so would break up the narrative. Those people who wish to investigate the finer detail for themselves will find the following papers a helpful starting point:

» Vilborg, L. 1955. The uplift of stones by frost. *Geografiska Annaler* 37(3/4): 164–169.
» Mackay, J. & Burrous, C. 1979. Uplift of objects by an upfreezing ice surface. *Can. Geotech. J.* 16: 609–613.
» Gleick, J. 1987. Sometimes heavier objects go to the top: here's why. *The New York Times*, 24 March.

» Luckman, B. 2013. Mountain and hillslope geomorphology. In: Schroder, J. (ed.). 2013. *Treatise on Geomorphology*, pp. 174–182. Elsevier, London.

» Rixhon, G. & Demoulin, A. 2013. Glacial and periglacial geomorphology. In: Schroder, J. (ed.). 2013. *Treatise on Geomorphology*, volume 8, pp. 392–415. Elsevier, London.

The simple explanation is, however, as follows:

Rainwater sinks through the clasts to the rock face beneath. As it washes it takes the dust particles down and they fill the gaps. Water flowing down the face also washes soil and dust particles down. Frost 'up-freezes' the wet material from the solid rock bed beneath, and up to the surface. This water expands in the soil, pushing the loose rock surface above up like a swelling. This creates gaps between the clasts. The bigger clasts leave bigger gaps. The smaller clasts present among the bigger clasts fall into the gaps. The frost melts and the loose rock surface settles. The more winters pass in this way without the surface layer being disturbed, the more advanced the segregation becomes from smaller to larger, from the solid rock bed up to the loose rock surface. This is why the largest clasts are on the surface, and the deeper you go the smaller the clasts and the more 'solid' the material becomes.

The point is that if you find a scree with a pebble or chip surface, it is improbable that this hides a deep layer of cobbles and plates over boulders and blocks. And in fact, what tends to happen is that plates and blocks rise to the surface and stay there because they are two dimensional and rest on the rounded material beneath.

In a pebble or chip scree the clasts are all similar sizes. As a result, the landform is perpetually mobile and the clasts circulate like clothes in a tumble dryer (although at a rate that is too slow to be seen except by a time-lapse camera taking photographs over decades). On a scree with a mix of pebbles, chips, cobbles and plates, the pebbles and chips may circulate and the cobbles may jostle but any plates will rise to the surface and remain there. Boulders and blocks will move down the face but not sink into it. As a result, plates on a scree surface may house lichen communities of significant ages, which is less common on cobble surfaces and simply does not happen on pebble or chip scree.

Nevertheless, the higher the altitude the more active the congelifraction is and the more pebbles and chips are falling with the result that they flow down and become intermingled with patches of cobbles and plates, and out into wider blockfields.

With that in mind, it may be easier to move away from the position a tagged bat is roosting in, and simply perform a small-scale investigation elsewhere on the landform to get an idea of the likely maximum depth the bat has penetrated.

With all the information above recorded on a form, the record can be compared with the bats species summaries. However, before we rush on it is worth identifying 'when' bats have been recorded in loose rock. The season in which the Daubenton's bat recorded by Altringham (2003) occupied scree is not given. The Natterer's bat recorded by Billington (2000) was recorded in July. The whiskered bat recorded by Michaelsen *et al.* (2013) occupied its scree in October and November. The brown long-eared bat recorded by Michaelsen & Grimstad (2008) appears to have occupied the scree roost in late September, and the other recorded by Michaelsen *et al.* (2013) was observed in early October.

4.5 The loose rock bat species summary

Unfortunately, only the scree within which Geoff Billington recorded the Natterer's bat is described in sufficient detail to be able to construct a bat species summary. However, this does offer some insight and visible cues that may help people prepare for a day out by programming their semantic memory.

4.5.1 Natterer's bat *Myotis nattereri* loose rock roosting account

The Natterer's bat loose rock face roosting account is constructed from Geoff Billington's own record and a later visit made to characterise the scree and situation. This record is summarised below and a photographic account is provided in Figure 4.10.

Table 4.2 Natterer's bat *Myotis nattereri* loose rock roosting account.

NUMBER OF ROOSTS FROM WHICH THE SUMMARY IS CONSTRUCTED					1	
ROCK TYPE	Sandstone					
SEASONS OCCUPIED	Winter: ?		Spring-flux: ?		Pregnancy: ?	
	Nursery: ✓		Mating: ?		Autumn-flux: ?	
	The number of records of Natterer's bat roosts in loose rock occupied in each season of the year					
LANDFORM OCCUPIED	Scree: ✓			Blockfield: ?		
Landform composition – Proportion of surface occupied by	**Pebbles:** 5%	**Chips:** 5%	**Cobbles:** 0	**Plates:** 85%	**Boulders:** 0	**Blocks:** 5%
ASPECTS OF FACES HOLDING ROOSTS	W					
ANGLE OF SLOPE	35°					
CONTIGUOUSNESS	Discrete in a wider network of patches					
HABITAT AROUND AND ON THE LANDFORM	A1.1.1 – Broadleaved semi-natural woodland; C1.1 – Continuous bracken; and, J5 – Other habitat / gorse					
% ISLAND VEGETATION PRESENT	5%					
% MOSS COVER	30%					
% CLASTS WITHOUT MOSS COVER THAT HOLD LICHENS	100%					
ENTRY POINTS	Superabundance					

Figure 4.10 The scree in which Billington (2000) recorded a Natterer's bat *Myotis nattereri* roosting. Top left: contextual satellite view. Top right: the view of the scree from the bottom on the woodland edge looking up. Bottom: the situation the bat was actually roosting in. (Top left image: © Google Images)

4.6 The point

Billington (2000) was convinced that when the Natterer's bat he tagged emerged from its scree roost, others emerged with it from other points over the scree face. Tore Christian Michaelsen is convinced that mountain scree in Norway is more widely used as a roost site by whiskered bats and brown long-eared bats than he has been able to prove to date. The reports of eight bat species crawling into subterranean rubble, suggest that loose rock might have a wide appeal.

At the moment, radio-tracking studies have no central online repository in which individual records might be stored to build a body of evidence.

At the moment, we do not even know how widespread screes and blockfields are as a lowland resource.

If you complete the recording form, you will have a priceless piece of data whether there is a bat in among the clasts that day or not. The simple act of recording the landform will make you look at it more critically and programme it into your mind so you will begin to see subtle differences in each new example.

And if you submit it to the *BRHK Rock Face Database* you will advance scientific knowledge and leave evidence that your life was meaningful and worthwhile long after you cease to exist.

4.7 Just for interest

The literature review of the use of loose rock by bats was disappointing: although the enthusiasm of the Norwegian team is palpable in their papers, the overall paucity of records was deflating. The fact is that without the roost description provided by Geoff Billington there was not really enough substance to justify this chapter. However, what did strike us was just how much work there has been on invertebrates. Some specialist invertebrates rely on scree as a development environment, and many other species may seek temporary shelter there. In addition to four bird species there were: 40 spiders; a pseudoscorpion; seven harvestmen; eight centipedes; six millipedes; three woodlice; 13 snails; nine beetles; a seed bug; a snail-killing fly; two bees; three moths and a butterfly. This likely represents a small fraction of the species that might be encountered within scree, although sufficient detail is rarely provided to confirm that a species has been recorded in scree rather than among rocky habitats, or under stones embedded in soil. In this respect, work by National Museums Liverpool (2018) in Snowdonia specifying records from scree was particularly valuable. Accepting that the bats are unlikely to take the snails, the list suggests that scree might be the equivalent of an all-you-can-eat buffet to a bat. For interested parties, the results are provided in Table 4.3; the seasonal period for invertebrates represents published periods of adult activity when these are most recognisable, though species developing among scree will be present as immature stages at other times of year.

Table 4.3 The organisms that are known exploit loose rock in either coastal or inland situations, and the period in which they are usually nesting (birds) or recognisable as adults (invertebrates), compiled using Sharrock (1976); Kerney (1999); Harvey *et al.* (2002a,b); National Museums Liverpool (2018); and, Gallon (2020).

SCIENTIFIC	COASTAL	INLAND	MONTHLY OCCUPATION
Leach's petrel *Oceanodroma leucorhoa*	✓	X	May–Aug
Puffin *Fratercula arctica*	✓	X	Apr–Aug
Wheatear *Oenanthe oenanthe*	?	✓	Apr–Jun
Snow bunting *Plectrophenax nivalis*	X	✓	May–Jun
Funnel-web spider *Tegenaria silvestris*	?	✓	Jan–Dec
Toothed weaver spider *Tetrix denticulata*	?	✓	Jan–Dec
Hacklemesh weaver spider *Coelotes atropos*	?	✓	Jan–Dec
Ground spider *Drassodes cupreus*	?	✓	Apr–Dec
Ground spider *Drassodes lapidosus*	?	✓	Feb–Dec
Ground spider *Zelotes apricorum*	?	✓	Feb–Nov
Mesh-web weaver *Cryphoeca silvicola*	?	✓	Jan–Dec
Dwarf sheet spider *Iberina montana*	?	✓	Jan–Dec
Money spider *Agyneta gulosa*	?	✓	Mar–Oct
Money spider *Centromerita concinna*	?	✓	Jan–Dec
Money spider *Centromerus prudens*	?	✓	Jan–Dec
Money spider *Diplocephalus cristatus*	?	✓	Jan–Dec
Money spider *Erigone arctica*	?	✓	Jan–Dec
Money spider *Micrargus herbigradus*	?	✓	Jan–Dec
Money spider *Microneta viaria*	?	✓	Jan–Dec
Money spider *Neriene peltata*	?	✓	Jan–Dec
Money spider *Obscuriphantes obscurus*	?	✓	Mar–Dec
Money spider *Oreoneta frigida*	?	✓	Jan–Nov
Money spider *Palliduphantes ericaeus*	?	✓	Jan–Dec
Money spider *Pocadicnemis pumila*	?	✓	Jan–Dec
Money spider *Poeciloneta variegata*	?	✓	Jan–Dec
Money spider *Porrhomma montanum*	?	✓	Jan–Nov
Money spider *Saaristoa abnormis*	?	✓	Jan–Dec
Money spider *Tenuiphantes tenebricola*	?	✓	Jan–Dec
Money spider *Tenuiphantes tenuis*	?	✓	Jan–Dec
Money spider *Tenuiphantes zimmermanni*	?	✓	Jan–Dec
Wolf spider *Pardosa agrestis*	?	✓	Apr–Sep
Wolf spider *Pardosa pullata*	?	✓	Mar–Nov
Wolf spider *Pardosa trailli*	?	✓	May–Aug
Wolf spider *Trochosa terricola*	?	✓	Feb–Nov
Goblin spider *Oonops pulcher*	?	✓	Jan–Dec
Jumping spider *Neon reticulatus*	?	✓	Jan–Dec
Jumping spider *Neon robustus*	?	✓	Mar–Oct

SCIENTIFIC	COASTAL	INLAND	MONTHLY OCCUPATION
Jumping spider *Pseudeuophrys erratica*	?	✓	Feb–Oct
Comb-footed spider *Robertus lividus*	?	✓	Jan–Dec
Comb-footed spider *Theonoe minutissima*	?	✓	Jan–Dec
Comb-footed spider *Rugathodes bellicosus*	?	✓	Mar–Nov
Crab spider *Ozyptila atomaria*	?	✓	Jan–Dec
Snake's-back tube-weaver *Segestria senoculata*	?	✓	Jan–Dec
Spurred orb-weaver *Trogloneta granulum*	?	✓	Mar, Oct
Pseudoscorpion *Neobisium carcinoides*	?	✓	?
Harvestman *Mitostoma chrysomelas*	?	✓	Jan–Dec
Harvestman *Nemastoma bimaculatum*	?	✓	Jan–Dec
Harvestman *Leiobunum blackwalli*	?	✓	Jan–Dec
Harvestman *Megabunus diadema*	?	✓	Mar–Jun
Harvestman *Mitopus morio*	?	✓	May–Nov
Harvestman *Nelima gothica*	?	✓	May–Dec
Harvestman *Opilio canestrinii*	?	✓	Jun–Dec
Centipede *Geophilus flavus*	?	✓	?
Centipede *Geophilus easoni*	?	✓	?
Centipede *Geophilus insculptus*	?	✓	?
Centipede *Lithobius borealis*	?	✓	?
Centipede *Lithobius calcaratus*	?	✓	?
Centipede *Lithobius crassipes*	?	✓	?
Centipede *Lithobius forficatus*	?	✓	?
Centipede *Lithobius variegatus*	?	✓	?
Millipede *Nanogona polydesmoides*	?	✓	?
Millipede *Chordeuma proximum*	?	✓	?
Millipede *Glomeris marginata*	?	✓	?
Millipede *Ophyiulus pilosus*	?	✓	?
Millipede *Proteroiulus fuscus*	?	✓	?
Millipede *Tachypodoiulus niger*	?	✓	?
Pill woodlouse *Armadillidium vulgare*	?	✓	Jan–Dec
Common shiny woodlouse *Oniscus asellus*	?	✓	Jan–Dec
Painted woodlouse *Porcellio spinicornis*	?	✓	Jan–Dec
Large chrysalis snail *Abida secale*	?	✓	?
Plaited door snail *Cochlodina laminata*	?	✓	?
Craven door snail *Clausilia dubia*	?	✓	?
Common chrysalis snail *Lauria cylindracea*	?	✓	?
Dwarf snail *Punctum pygmaeum*	?	✓	?
Moss chrysalis snail *Pupilla muscorum*	?	✓	?
Rock snail *Pyramidula pusilla*	?	✓	?
British whorl snail *Truncatellina callicratis*	✓	X	?

SCIENTIFIC	COASTAL	INLAND	MONTHLY OCCUPATION
Mountain whorl snail *Vertigo alpestris*	X	✓	?
Common whorl snail *Vertigo pygmaea*	?	✓	?
Milky crystal snail *Vitrea contracta*	?	✓	?
Ribbed grass snail *Vallonia costata*	?	✓	?
Glass snail *Vitrea subrimata*	X	✓	?
Ground beetle *Abax parallelepipedus*	?	✓	May–Sep
Ground beetle *Badister bullatus*	?	✓	Apr–Aug
Ground beetle *Harpalus latus*	?	✓	May–Sep
Ground beetle *Leistus montanus*	?	✓	Jul–Sep
Ground beetle *Leistus spinibarbis*	?	✓	Jun–Sep
Ground beetle *Olisthopus rotundatus*	?	✓	May–Oct
Ground beetle *Pterostichus madidus*	?	✓	Jul–Sep
Black snail beetle *Silpha atrata*	?	✓	?
Rove beetle *Tasgius melanarius*	?	✓	?
Seed bug *Drymus brunneus*	?	✓	?
Snail-killing fly *Ectinocera borealis*	?	✓	May–Jul
Fabricius' nomad bee *Nomada fabriciana*	?	✓	Mar–Aug
Common furrow-bee *Lasioglossum calceatum*	?	✓	Mar–Oct
Green carpet *Colostygia pectinataria*	?	✓	Apr–Oct
Grey mountain carpet *Entephria caesiata*	?	✓	May–Oct
Northern spinach *Eulithis populata*	?	✓	Jun–Sep
Small heath *Coenonympha pamphilus*	?	✓	May–Sep

Subterranean rock: characterising and recording the landforms and the Potential Roost Features they may hold

Henry Andrews

In this chapter	
Preamble	Recap of the previous chapter and relevant project objective; how Chapter 5 dovetails into Chapters 2, 3 and 4 and contributes to achieving the project objectives; and, a recap of the component parts of a biological record – who, where, when and what
An introduction to the subterranean landforms	The four landforms – solution caves, sea caves, worked-out mines, and decommissioned railway tunnels; introducing solution caves; introducing sea caves; introducing mines; introducing railway tunnels
An introduction to 'what' to record in the context of subterranean landforms	The three recording forms; introducing the subterranean bat species; a summary of the months in which the individual species have been recorded underground; introducing the 'What 1' macro-environment values; introducing the 'What 2.1' meso-environment values; introducing the 'what 2.2, 2.3 and 2.4' micro-environment and roost account
What 1: recording the subterranean rock macro-environment values	The rock type; the landform; the entrance topography; the habitat on/over the subterranean rock landform entrance(s); the number of 'bat-friendly' entrances; water
What 2.1: recording the subterranean rock meso-environment values	Meso-environment 1a – Roost situation and illuminance; Meso-environment 1b – Roost distance from the entrance; Meso-environment 2 – Roost position; Meso-environment 3 – Routes, junctions and destinations; Meso 4a – Horizontal routes; Meso 4b – Vertical routes; Meso 4c – Enlarged nodal and destination features
What 2.2: recording the subterranean rock micro-environment values for exposed plane surface and recessed roost positions	Micro 1 – The bat roost record for exposed plane surface situations; Micro 2a – The bat roost record for recessed situations; Micro 2b – The bat roost record for double-recessed situations; Micro 3 – The associate species
What 2.3: recording the subterranean rock micro-environment values for enveloping roost positions	Micro-environment 1 – Enveloping PRF in plane surface situations; Micro-environment 2 – Enveloping PRF inside recesses; Micro-environment 3 – The enveloping PRF itself

What 2.4: recording the advanced values for enveloping roost positions	The distance the bat is from the entrance; the internal dimensions; whether a comprehensive inspection was possible; internal humidity; associated species
The subterranean climate	Introducing the invisible niche; what we need to know to use the knowledge to inform surveys of new sites and monitoring of artificial landforms; airflow; temperature; humidity; illuminance
The subterranean rock bat species summaries	Barbastelle *Barbastella barbastellus*; serotine *Eptesicus serotinus* Bechstein's bat *Myotis bechsteinii*; Brandt's bat *Myotis brandtii*; Daubenton's bat *Myotis daubentonii*; whiskered bat *Myotis mystacinus*; Natterer's bat *Myotis nattereri*; common pipistrelle *Pipistrellus pipistrellus*; brown long-eared bat *Plecotus auritus*; grey long-eared bat *Plecotus austriacus*; lesser horseshoe bat *Rhinolophus hipposideros*; greater horseshoe bat *Rhinolophus ferrumequinum*

WARNING

This chapter has been written to inform ecologists who:

1. Are under the supervision and instruction of people who are competent and qualified to enter caves, mines and tunnels and are familiar with the layout and hazards of the specific system that is to be surveyed;

2. Have had sight of an up-to-date engineer's report that confirms the cave, mine or tunnel is safe to enter;

3. Have completed an appropriate Risk Assessment in line with the minimum requirements as set out by the HSE;

4. Are appropriately equipped;

5. Are aware of what is expected of them in an emergency, and competent to perform all tasks allocated to them.

This chapter is absolutely NOT an encouragement for anyone who is not appropriately competent and qualified to enter any subterranean environment or even into the near vicinity of that environment: remember, the hazards may not all be underground.

5.1 Preamble

In Chapter 2 the different rocks the reader might encounter in their part of the British Isles were identified. The landforms made of those rocks were then described so that they might be identified in the field. In Chapter 3 the landforms that offer surface rock faces were described in the context of bat roost habitat, and the same was done in Chapter 4 with loose rock landforms. Although semi-subterranean situations may be offered by talus, the truly subterranean environment is only offered by caves, mines and tunnels. And so, at the start of this chapter we find ourselves metaphorically standing at the mouth of the final group of natural and semi-natural rock landforms.

As with rock faces and loose rock landforms, we begin the satisfaction of Project Objective 2 – 'the collation of published descriptions of bat roosting ecology on, among and in natural and semi-natural rock habitats' – by introducing the subterranean landforms.

And just as we did in Chapters 3 and 4, we apply the rationale set out in Chapter 1, using the standardised *BRHK Subterranean Rock Database* recording form as the framework for the data review and the chapter layout. This will mean that the information that has been collated is presented in a format that has a practical application.

To recap, in Chapter 1 we looked at the 'who', 'where' and 'when' of a biological record. Among this information, the 'where' allows us to identify the macro-environment in terms of the climate and geology.

This chapter comprises 'what' else we need to record for a useful record of a specific subterranean landform. Obviously, if an inspection of a PRF encounters bats and/or droppings this immediately confirms it is a roost, but what if the PRF is empty: does that mean it is not suitable to hold roosting bats? Answering this question is achieved by a comparison of environmental variables of the empty PRF against those of known roosts.

This chapter describes the variables that we need to record for comparison, and means that at the end of a day where no bats have been encountered, we will nevertheless have a useful body of data that will mean knowledge of the sort of PRF different rock types can offer will build over time to a point where a reliable confidence threshold is identified for the benefit of: **a)** the bats themselves; **b)** the surveyor; and, **c)** any developer that wants to have a go at building an artificial landform that will have a tangible value to bats, and that the developer can therefore offer as part of a strategy to deliver biodiversity net gains.

This chapter sets out the evidence that might be used to assess whether or not the empty PRF is suitable for a particular bat species to exploit it as a roost, and whether it is in fact a roost already.

5.2 An introduction to the subterranean landforms

In the context of the BRHK project, the subterranean rock landforms comprise: **1)** solution caves; **2)** sea caves; **3)** worked-out mines; and, **4)** decommissioned railway tunnels.

> **Note:** In the natural caves only solution and sea caves are considered. Talus cavelets were identified in Chapter 4 and will not be considered further in this chapter. Furthermore, while they are unquestionably of significant importance to bats in Eastern Europe, the BRHK project specifically does not consider military bunkers or silos. This is because they are not landforms that are commonly encountered in the British Isles in anything like the size and complexity that would make them anything other than atypical.

To set the scene, the four landforms are illustrated in Figures 5.1–5.5.

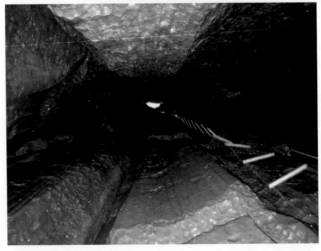

Figure 5.1 The characteristically vertical entrance to a natural solution cave in the north of England. (All images: Rich Flight – Flight Ecology)

Figure 5.2 Characteristically horizontal entrances into three natural solution caves in the south of England. Top: Oak Cave has an entrance sufficiently large to drive a Ford Transit in and turn it round. Middle: Goatchurch Cavern is a walk-in. Bottom: Big Cave is named because it is just large enough for one caver to squeeze into.

Figure 5.3 The entrances to three natural sea caves. Top: some caves are only accessible over water, but if they do not flood at high tide their high ceilings still offer roost opportunities. Middle: some are not unlike solution caves. Bottom: some caves are narrow and offer a network of crevices in sunlit situations.

Figure 5.4 The complexity and scale of mines. Top: it can take a day just to survey the entrance. Middle: some mines are so complex that surveying needs careful planning if areas are not to be missed. Bottom: some mines are just so large that the bats could be on an exposed ceiling and still be overlooked. (All images: Stuart Spray Wildlife)

Figure 5.5 Top: some railway tunnels are sanitised and illuminated. Middle: some railway tunnels are dark but disturbed at night. Bottom: and some railway tunnels are just right.

5.2.1 Introducing solution caves

The definition of a cave is simple: a natural opening, usually in rock, that is large enough for a person to enter (Waltham *et al.* 1997; Klimchouk 2004). A cave might be air or water-filled, but the length or height or depth will be greater than the cross-section dimension at the entrance (Klimchouk 2004).

Caves may be formed by different mechanisms depending upon the rock, and the UK has examples of simple inland caves in granite and sandstone. However, inland natural cave systems are most common in limestone (including chalk) and, less often, dolomite,[21] within a karst[22] landscape (Vandel 1965; Waltham *et al.* 1997).

As the name implies, solution caves are formed by solution by water that is high in carbonic acid, which dissolves the rock. As this relies on an increased concentration by the production of carbon dioxide by plant roots, the development of caves is maximised in hot, wet, tropical environments (Waltham *et al.* 1997). Although our climate is cool now, it was once much warmer and the caves we see today were formed in prehistoric times. The solution caves that remain today are sections of bigger networks, essentially truncated fragments of the larger conduits of the prehistoric karst drainage net (Poulson & White 1969).

Waltham *et al.* (1997) describe three stages in the life of a cave passage, comprising:

1. **Initiation**, which creates the entrance fissures in the rock and allows the flow of groundwater;

2. **Enlargement**, where the initial fissures are progressively increased in diameter over thousands of years (N.B. it may take 10,000 years to reach 1 m in diameter (Mylroie & Carew 1987)). This continues until the upper zone drains and becomes vadose and the saturation zone moves progressively lower, and what was typically a complex beginning with many entry points, gradually becomes an increasingly simpler network;

3. **Degradation**, where the individual passages ultimately: **a)** collapse; **b)** are clogged by the inflow of sediment; or, **c)** sections of the system are worn away over geological time. N.B. Although inflow of sediment is common in the British Isles, collapses are uncommon and the wearing down of strata to break up and even erase subterranean systems would take millennia to occur.

In the British Isles a passage 5 m in diameter would be considered large, but there are some that exceed 15 m (Waltham *et al.* 1997). In terms of extent, two in the British Isles exceed 50 km; Ease Gill and Ogof Ffynnon Ddu, the latter of which is also the deepest at 308 m (Waltham *et al.* 1997).

The simplest caves consist of only one element (such as a passage, or individual cavern) but these are unusual and most are networks of caverns and chambers linked by avens and passages (Klimchouk 2004). Klimchouk (2004) lists four recognised patterns of cave, comprising: **1)** single-conduit; **2)** single-void; **3)** branchwork; and, **4)** maze.

Single-conduit and single-void caves are the result of a single stream of water, but the water arrives from different directions. In a single-conduit the water is flowing from the surface down, but a single-void cave is more often the result of hypogenic flow, i.e. with the water rising up from below.

The same division applies to branchwork and maze caves. Branchwork caves are the result of converging tributaries arriving from several points above the surface and are the most common cave pattern (Palmer & Audra 2004). Maze-caves encompass closed loops in which the passages have all formed simultaneously as the result of the flow percolating upward through a network of fissures in which no one fissure allowed a significantly greater flow than any other (Klimchouk 2000; Palmer & Audra 2004). In the

British Isles nine extensive maze caves are known in the Pennines, of which all but one was discovered in the nineteenth century by lead miners (Farrant & Harrison 2017). Some of these are truly vast, extending over 1 km in length and with passages up to 5 m high and 2–3 m wide in patterns that criss-cross the maps in eye-twisting contortions (see Farrant & Harrison 2017).

Regardless of whether the cave is an epigenic single-conduit or branchwork cave, or a hypogenic single-void or maze cave, all caves are strongly influenced by the nature of the surface landscape, and in particular rivers and streams (Cullingford 1962).

Moving on, caves may be further classified by their orientation into three types, comprising: **1)** vertical; **2)** inclined; and, **3)** horizontal (Klimchouk 2004) and each individual system can be divided into three zones, comprising: **a)** the 'vadose zone' in which the water table has declined leaving the cave more-or-less dry and only damp underfoot; **b)** the 'epiphreatic zone' within which the water table appears and flows over the floor, and which may also be seasonally flooded; and, **c)** the deep 'phreatic zone' which is permanently flooded (Littva *et al.* 2017).

Large-diameter passages have typically been formed when they were phreatic, but this may continue in the epiphreatic zone with solution taking place as the result of water flowing along the floor. Phreatic passages that are becoming vadose may be altered from a simple rounded tube to become keyhole-shaped as a result of water flowing along the floor, moving pebbles with it, and cutting down a canyon channel (see Mihevc *et al.* 2004; Sparrow 2009).

These, often complex systems, are the only erosional environment where each subsequent event does not alter those that have already occurred above (Waltham *et al.* 1997). This is because the erosion results in downcutting of conduits at ever lower strata and the passage sections above become redundant as the erosion continues ever deeper underground. This leaves the deposited boulders, cobbles, pebbles and silts for palaeontologists to search through.

Although the largest area of karst is on the Cretaceous Chalk, this landscape is limited in cave development and the greatest proportion of caves occur within the limestone of the Lower Carboniferous (Waltham *et al.* 1997). As a result, even where caves are common in a locality, they tend to be clustered in a relatively small area, or spread out at intervals along linear geological elements. In the British Isles limestone caves are primarily aggregated in hill country to the north and west of a line running from the mouth of the Tees to the mouth of the Exe (Chapman 1993). Regardless, as a bat roost habitat resource they are by no means common to all counties in the British Isles, and in some counties they are entirely absent.

The six primary cave-rich areas in the British Isles are listed by Chapman (1993) and Waltham *et al.* (1997) as: **1)** County Fermanagh (Northern Ireland); **2)** The Pennines (County Durham, Northumberland, North Yorkshire & Cumbria); **3)** The Peak District (Derbyshire); **4)** County Clare (Republic of Ireland); **5)** South Wales (principally Carmarthenshire, Powys & Monmouthshire); and, **6)** The Mendip Hills (Somerset). A map showing the counties in which solution caves do certainly exist as well as those in which they might occur within karst exposures is provided in Figure 5.6.

These caves are for the most part in carboniferous limestone, but their characters differ according to their geological planes: the planes in the north are the right way up (although fragmented by vertical faults); the planes in the south tend to exhibit the signs of buckling with steeply dipping bedding planes (Chapman 1993). As a result, the northern cave systems tend to follow a stepped system of 'potholes' with vertical shafts, dropping to horizontal passages and on down another shaft, etc. Where vertical faults are absent, the cave may be more-or-less horizontal (Chapman 1993). In contrast,

Figure 5.6 The counties that hold caving areas and might also offer solution features in the karst landscape of the British Isles.

the caves of the Mendip Hills have steep passages, but few descending vertical sections (Chapman 1993). Notwithstanding, some have ramps with a step up at the base which again gives way to a ramp, and another step (op. cit.). As a result of the broad differences from north to south, a northern spelunker will speak of potholing, and a southern spelunker will speak of caving (Cullingford 1962).

Away from these broad differences, the systems may commonly encompass: **a)** large open caverns and smaller chambers; **b)** deep fissures in the ceilings, walls and floor; **c)** boulders, cobbles, pebbles and silts; and, **d)** a host of speleothems, such as stalactites, stalagmites and curtains.

The ceilings, walls and floor also encompass a wide range of dissolutional forms (Mihevc *et al.* 2004). Features such as flutes, scallops, domes, spires and pendants may be encountered in and on ceilings, as well as flutes and scallops on walls. Potholes formed by spun rocks may be encountered in the cave floor (Mihevc *et al.* 2004). Breakdown is another common feature of solution caves and comprises rock that has fallen from the cave ceiling, leaving the ceiling domed as a result (Mihevc *et al.* 2004). Breakdown is effectively subterranean talus (see Chapter 4).

Note: The introduction above takes into account recent advances in knowledge. Notwithstanding, there are no better books than:

- Cullingford, C. (ed.). 1962. *British Caving: An Introduction To Speleology*, 2nd edn. Routledge & Kegan Paul, London
- Sparrow, A. 2009. *The Complete Caving Manual*. Crowood Press, Wiltshire

Anyone who is intending to visit a cave should read both books from cover to cover first, and implement all the advice given by Sparrow (2009).

5.2.2 Introducing sea caves

In the context of the BRHK project, sea caves include any system extending back into an escarpment from the littoral zone (i.e. the beach). Sea caves may form in rock by a process of wave action with associated abrasion (Klimchouk 2004).

Moore (1954) describes the specific conditions for sea cave formation as: **1)** an escarpment that is in contact with the erosive force of waves and currents; **2)** a face on the escarpment that encompasses specific geological structures or textures that allow differential erosion; and, **3)** comprises rock that is sufficiently resistant to prevent the formation of a protective beach.

The cave comes about as a result of the cliff against which the waves break having a zone of relative weakness into which the wave action can erode a fissure, which is widened as the result of the force of the sand loaded wave being in a confined space (Bunnell 2004). As the wave action is limited by the reach of the tide, sea caves are shorter than solution caves and tend to be simple passages (Bunnell 2004). However, whether the cave is simple or complex will depend upon the strength and direction of the waves and currents, coupled with the geology and slope of the seabed (Moore 1954).

Note: True sea caves have no element of dissolution, i.e. they are not solution caves that open into the sea (Bunnell 2004). This separation is not pertinent to the BRHK project, and we will consider any cave that opens onto a beach as a sea cave regardless of the mechanism that made it.

Sea caves occur most frequently on coasts comprising rocks that are strong enough to stand in near-vertical faces and to support a roof (Woodroffe 2003). In the UK sea caves have a northern and western bias (Chamberlain 2004).

Unlike solution caves which are complex internally, sea caves are generally (although not always) simple, opening with a wide mouth and with the walls tapering in at the apex of the roof and the floor rising up towards the back. However, ceiling erosion is common due to the wave and air pressure and typically results in a high peak tapering into a fissure. In larger sea caves this may result in collapses and a wide arch forming.

A compilation of sea caves around the world with lengths of over 100 m has been compiled by Bob Gulden and is available at www.caverbob.com/seacave. This list includes 24 that occur in the UK, of which 18 (75%) occur in Scotland and the remaining six (25%) occur in England.[23] The Scottish caves range in length from the 107 m of Mackinnon's Cave on the Isle of Staffa, to the jaw-dropping 230 m of Sandside Head Cave at Thurso (ranked 26th longest cave in the world). The English caves range from 100 m at Livermore Head in Devon, up to 225 m of Virgin's Spring under the island of Lundy, which lies off the north coast of Devon.

5.2.3 Introducing mines

Mines have been sunk since humans first appeared in Britain. Edwards (2011) cites our ancestors sinking pits up to about 15 m deep 5,000 years ago. After this (despite only patchy evidence) the Romans get a good deal of credit from a wide range of sources for opening many of the limestone and coal mines that were later extended. Following on, the Norman conquest is suggested to have been a time of increased mining, but the written record only appears to begin properly in the 1300s.

Notwithstanding, although almost all Britain's coalfields were being worked by the 1500s (Hayman 2016), a clear narrative is only available from the mid-1600s, around 100 years after a decline in mining brought about by the cessation of church building following the dissolution of the monasteries.

Deep mining appears to take off again during the Victorian era, when steam engines made the pumping of water possible. After that we see a boom between the wars and a peak in coal mining from over 1,500 collieries in the 1940s (Hayman 2016), before the deep mines begin to close and activity switches to the opencast method, with deep mining dwindling until all are closed by 2016.

Regardless of their start date, the greater proportion of the extent of mines that remain as subterranean systems appear to have been dug in the nineteenth and twentieth centuries.

Some of the earliest shafts are impressively deep: in the eighteenth century they were up to 70 m deep and pillar-and-stall mining might extend 140 m out from the base (Hayman 2016). These depths and extents were soon increased and the internal extents can be jaw-dropping in their scale. For example, some mines worked in the Victorian era were extended almost as far as natural cave systems and encompass caverns with ceilings 15–25 m above the floor (Jones 2003, Edwards 2011). The pyramidal-shaped 'Echo Chamber' within Penn Recca Slate Mine has a floor 12 x 24 m ascending to a peak some 27 m above (Edwards 2011), and the 'Great Chamber' within Lamb Leer lead mine has a floor diameter of c.20 m and over 30 m in height (Burr 2015a). The end of the nineteenth century appears to have seen the maximum extents, with some coal faces over a mile from the shafts that serviced them (Hayman 2016).

Just as natural cave systems are restricted to karst so the geography of mine working is limited by the geology. In the UK, deep mines that offer the same environments as natural systems include (but are not restricted to): limestone, ochre, slate, coal, chalk, calcium, tin, copper, iron, zinc, lead, silver, manganese, strontium, and barium.

Mine landforms encompass: **1)** simple 'bell-pits'[24] with a wide diameter shaft dropped and then adits that were little more than crawl spaces extended off from the base; **2)** individual 'dead-ended' horizontal adits that follow single seams; **3)** vertical shafts that drop to longwall[25] coal faces or ball clay 'drive and fan';[26] **4)** multilevelled networks that follow seams of saleable material while also seeking drainage outflow channels; and, **5)** more-or-less level 'pillar-and-stall'[27] labyrinths in limestone strata and along coal seams.

The vertical shafts remaining from the nineteenth century appear to be of either square or rectangular shape (known as the 'Cornish' standard (Burr 2015a)). The square shafts are typically 1.5–1.8 m² (Burr 2015b). The sizes of rectangular shafts vary but 1.8 x 2.4 m dimensions are common (Burr 2015a). The upper section of the shaft was typically lined with timber to stabilise earth and loose material but as they descended into stone the lining was no longer needed. In the nineteenth century drops up to 30 m deep were common, anything up to 60 m not unusual and some descended over 70 m straight down[28] (see descriptions given by Edwards (2011); Clarke *et al.* (2012); and, Burr (2015a)). Modern deep mining appears to have been restricted to coal, with shafts up to 900 m deep sunk in the 1920s (Hayman 2016). Nonetheless, in a great many cases, as soon as a shaft was no longer needed it was capped. Where access was no longer required, the capping would typically be achieved by infilling with waste stone, or by creating a plug or plate of concrete. Where access might be required, steel doors might be constructed, or a grille.

Horizontal adits may be driven to give access to the seam or face, or to provide drainage, or ventilation. Adit dimensions are typically rectangular and wider than tall, 2.5–3 m wide and 2–2.5 m tall. They may be straight or meandering but generally go in and slightly upward in order that water can flow down and out. The exceptions are the ball clay mines of Devon which used incline adits that descend and had to be pumped out (Edwards 2011; Hayman 2016). One broad rule is that (aside of coal mines) adits tend to be significantly longer in than shafts go down.

Just like natural systems, some eighteenth-century mines have passages with irregular shapes and dimensions, varying in diameter as the miners followed seams with constricted crawls opening into walkable sections which may then narrow again (see photographs in Burr 2015a).

Combinations of shafts and adits combine to create systems that extend over 50 m under the surface joining chambers, caverns and lakes, the floors of which are interconnected, spanning individual platforms or by stages reaching the world above (Edwards 2011).

Overall, the descriptions of mine systems suggest that they have a far greater number of access openings than natural systems, and although they are not created by water flow, that does not mean they are dry: there may be inflow of rainwater and the mine itself may break through faults through which water flows, or simply work below the water table. As a result, many worked-out systems are now entirely water-filled and others that remain are subject to predictable annual flooding, and dangerous flash flooding.

Edwards (2011) suggests that rewarding projects remain in respect of locating and recording all the shafts and adits of workings in old systems. This is unsurprising when we consider that when the National Coal Board was formed in 1947 there were over 1,500 coal mines alone (Hayman 2016). The situation is so complex that an attempt to construct a distribution map of mine workings in the British Isles had to be abandoned due to the publication deadline; *Homo sapiens* appears to be as happy burrowing as *Talpa europaea*, with very few areas left undug.

Mine historians are entirely open in the gaps in their knowledge, even of specifically where mineshafts and adits open, and what lies beneath those that are known. This suggests that mines are the final frontier for bat discoveries, due to: **a)** patchy published descriptive accounts of what they encompass; **b)** patchy and poorly detailed mapping; **c)** many being in private ownership on private land and away from convenient footpaths and bridleways; and, **d)** because mines are so very VERY dangerous.

These factors also act as guards against casual exploration, and mines may be the least disturbed environment a bat might occupy. Where access is still provided by an open adit, all those capped shafts have resulted in EXACTLY the kinds of condition favoured by horseshoe bats, and even where the bats are known to exploit the adit, it is entirely possible that significant greater numbers are out of reach and out of view at the apex of the shaft.

Clarke *et al.* (2012) and Burr (2015a and 2015b) provide maps, photographs and summary descriptions of ochre mines in the Mendips that encompass depths of 50 m and more. Although the farthest distances from the entrances are typically under 80 m, the complex adits actually fork and circle and thereby comprise hundreds of metres of underground passages that connect small chambers and larger caverns.

As a wide range of narratives describe nineteenth-century mine openings now being under a woodland canopy the probability that they are exploited is tantalisingly high. Yet geologists and cavers alike have expressed concern over the lack of knowledge with Burr (2015a) noting that a great many ancient mine workings have disappeared within his lifetime as a result of the determined actions of local farmers.

5.2.4 Introducing railway tunnels

Because people that study railways are obsessed with cultural minutiae there are hundreds of pages supposedly devoted to railway engineering that have no specification information whatsoever. It is therefore recommended that any bat group measure their tunnels and submit the information so that there is some functional record available. In the end so much time was wasted in the review that the following was the best that could be achieved.

Railway tunnels are the 'poor cousin' of the subterranean landforms. While there are a multitude of definitions, the Wikipedia definition has been adopted as the most sensible, and it is as follows:

An underground passageway with no defined minimum length, though it may be considered to be at least twice as long as wide. Some civic planners define a tunnel as 0.1 miles (160 m) in length or longer.

Wikipedia has dedicated lists of tunnels in the Scotland, Wales and England, and also Northern Ireland and the Republic of Ireland.

Of an overall 514 tunnels in Scotland, Wales and England, railway tunnels account for 412 (80%). Breaking this down by region, Scotland has 22 (6%), Wales has 46 (11%) and England has 344 (83%).

Of an overall 61 tunnels in Northern Ireland and the Republic of Ireland, railway tunnels account for 39 (64%). Breaking this down by region, Northern Ireland has 5 (13%) and the Republic of Ireland has 34 (87%).

The data are sufficient to produce a density map for the British Isles and this is provided in Figure 5.7.

Not all the Wikipedia entries have the date the tunnel was opened to traffic, but 339 do. Although a handful of canal tunnels can be dated to the 1700s,[29] the earliest date given for a railway tunnels is 1807 for the Middlebere Plateway Tunnel in Dorset. After this there is a pause before the engineers went mad in the 1840s and bored their way through 66 tunnels (16% of the total) in one year. In fact, all the tunnelling needed for

Figure 5.7 Density map of railway tunnels in the British Isles.

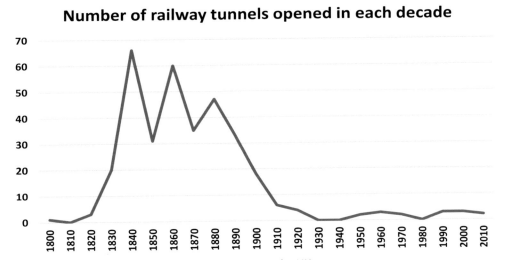

Figure 5.8 The peak of railway tunnelling activity in the UK.

the railway network was more or less put in place in the 100-year period from 1820 to 1920, as illustrated in Figure 5.8.

Pragnell (2016) provides a summary of tunnel construction, as follows: **1)** the route is surveyed over the hill; **2)** shafts are dug to extract the rock during construction and provide ventilation when the tunnel is in use; **3)** narrow adits are bored between the shafts and directed using a compass; **4)** when the adits are linked, the working gallery is increased to the desired height and the shape of the walls is profiled using a pre-shaped timber frame as a guide; **5)** drainage channels and soakaways are constructed to drain off incoming springs, etc.; **6)** the internal walls are lined by layers of brick or stone (two or more in thickness); and, **7)** the entrance portals are stabilised.

In terms of shape and dimensions, unless the tunnel is driven into an existing rock face, the line will approach in a cutting to reach the vertical face of the entrance arch. The shape of the arch usually indicates the shape and dimensions of the tunnel along the entire length. Pragnell (2016) describes arches ranging from semicircular to elliptical and horseshoe with the walls curving inward at the base.

The tunnels range in dimensions, but a typical single-track tunnel is approx. 4–6 m high and 6–7.5 m wide. As far as lengths go, if we concentrate on the period 1820 to 1920 and omit the London Underground, the shortest tunnel is the 27 m Peak Forest Tunnel which was opened in Derbyshire in 1863, and the range goes up to the 7 km Severn Tunnel which was opened between Monmouthshire and Gloucestershire in 1886. Figure 5.9 illustrates the proportion of tunnels in each 100 m group.

It was not possible to produce a meaningful figure for how many of these tunnels remain in use, but some decommissioned tunnels are already notable bat reserves, such as: **1)** Singleton and Cocking Tunnels Special Area of Conservation in West Sussex (notified for the number of barbastelle and Bechstein's bats that roost in the two tunnels, as well as the one greater mouse-eared bat left in England); **2)** Buckshraft Mine & Bradley Hill Railway Tunnel SSSI in Gloucestershire (notified for the numbers of lesser and greater horseshoe bats that roost in the two tunnels); **3)** the Malvern Hills SSSI which spans Herefordshire and Worcestershire (which is exploited by roosting lesser horseshoe bats); and, **4)** Withcall and South Willingham Tunnels SSSI in Lincolnshire (notified for the Brandt's bats, Daubenton's bats, Natterer's bats, whiskered bats and brown long-eared bats that roost there).

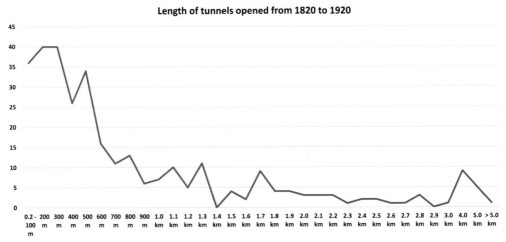

Figure 5.9 The number of railway tunnels in each 100 m range class.

Although at first sight the tunnels are effectively one single large-diameter tube, along the length springs were often encountered that required drainage, and if the tunnel was significantly long, it would also require ventilation shafts. The brick lining may be up to 70 cm thick to resist the seepage of water (Pragnell 2016), and that may result in a complex network of fissures as the mortar degrades. If we then factor in the need for refuge manholes along their lengths which are staggered on both sides (as we saw in Chapter 2), we have an artificial landform that is not entirely dissimilar to a cave system. Add in the value of these landforms to bats seeking a sheltered shortcut between foraging habitats and suddenly their probable appeal is glaringly obvious.

5.3 An introduction to 'what' to record in the context of subterranean landforms

The introductions now made, we may now move on to the 'what' of a biological record in the context of subterranean landforms.

The 'what' is all the meso- and micro-environment values.

The objective that we are attempting to secure is a series of designs for artificial roost landforms.

When recording the roost, it helps if you stand back and think about how you would instruct a builder to make it. Think about how you would describe the specification if you were talking to the builder on the phone:

> I need a 4 m long and 3 m wide chamber at the end of a 10 m long passage. The chamber will need a 2 m high ceiling. The ceiling will need a 1 m diameter domed cupola extending 30 cm up, and that cupola should have a 20 cm diameter pocket extending another 10 cm up into the apex of the cupola; basically, I want a miniature St Pauls Cathedral that I can bury and access along a tube … got it?

5.3.1 The three recording forms

Subterranean landforms have a macro-environment that is removed from that outside. This is not to say the subterranean landform is *disconnected* from the environment above ground, because the above-ground environment does penetrate and have a degree of influence throughout. Notwithstanding, the subterranean environment has a physical structure that encloses a large amount of space, and that sets solution caves, sea caves, mines and railway tunnels apart from rock faces and the loose rock landforms.

This removed environment has a narrower gradient in which the niche positions are easier to identify. But we first have to describe that environment, and ironically this is much harder underground than it is on the surface outside. In the subterranean landforms, we may have to travel significant distances actually inside the PRF itself.

Rock faces and loose rock landforms are the edge of our world. The subterranean landforms hold a world that may extend far beyond that edge.

Just as we find them in roof voids, some bat species may simply be exposed on the ceiling or a wall of a cave, or in a situation that is recessed from the primary surface where we could still take them with one hand. Other species may prefer an entirely enveloping situation where it is hard to imagine how they managed to gain entry. We do not encounter bats in exposed situations on rock faces or on loose rock. As a result, the rock face and loose rock recording forms are simpler than the subterranean landform recording form.

To anticipate even a basic record, the subterranean rock recording form is in three parts, as follows: **1)** the macro-environment; **2)** the meso- and micro-environment of exposed roost positions; and, **3)** the meso- and micro-environment of enveloping roost positions.[30] To refresh your memory, the *BRHK Subterranean Rock Database* recording forms are provided again in Tables 5.1–5.3.

Note: The macro values will not be recorded again. Subterranean rock landforms do not change in the way that faces and loose rock might, and unlike tree roosts (which are created and destroyed by storms, etc.) subterranean features are functionally permanent. To make the point; you might record a roost in a subterranean landform today that your great, great, great-grandchildren will go back and check in over 100 years' time; that is less likely to be true of a rock face and even less still with a loose rock landform.

Table 5.1 The *BRHK Subterranean Rock Database* recording form – macro values.

WHO	1st Recorder:	2nd Recorder:
WHERE	**Site name:** see Chapter 1	
	Grid reference of the primary entrance:* see Chapter 1	
WHEN	**Date:** see Chapter 1	
WHAT 1	**Macro – Rock type (e.g. granite):**	
	Macro – Landform: solution cave / sea cave / mine / railway tunnel / other:	
	Macro – Entrance topography: Does the landform open in from: a sheer face / sloping ground / level ground / a dell or pit / a cutting (tick as many as apply)	
	Macro – Habitat over the opening through which the surveyors entered (phase 1):	
	Macro – Number of 'bat-friendly' entrances (include shafts and chimneys/avens, etc.):	
	Macro – Does the landform hold a: free surface stream / sump stream / sump / lake / no water	

* The primary entrance is the one that offers the easiest access and the one that cave rescue would be most likely to use if they had to get you out, which is EXACTLY why you are recording it: in case you or anyone else gets into trouble!

Table 5.2 The *BRHK Subterranean Rock Database* recording form – meso and micro values for exposed surface and recessed roost positions.

WHO	1st Recorder:		2nd Recorder:	
WHERE	Site name: see Chapter 1			
WHEN	Date: see Chapter 1			

<table>
<tr><td rowspan="8">BASIC RECORD</td><td rowspan="7">WHAT 2.1</td><td colspan="2">Meso 1a – Is the roost in the: threshold / dark-zone</td><td colspan="2">Lux value:</td></tr>
<tr><td colspan="4">Meso 1b – What is the Distance between bat/PRF and nearest entrance (m):</td></tr>
<tr><td colspan="4">Meso 2 – Is the roost position/PRF on the: ceiling / left wall / right wall / face / floor</td></tr>
<tr><td colspan="4">Meso 3a – Is the roost feature: on a route / at a node or junction / in a destination
Now go to Meso 3b …</td></tr>
<tr><td colspan="2">Meso 3b – Is the roost feature on a route in a:
portal / aisle / passage / keyhole passage / adit / tunnel / crawl / squeeze   – if yes >> if no go to 3c</td><td colspan="2">What is the ceiling height (m):
What is the width between the walls (m):
Is the roost feature: in a wider section of a route (i.e. you can walk past it) / at the end of a cul-de-sac / in a face   Now go to Meso 4…</td></tr>
<tr><td colspan="2">Meso 3c – Is the roost feature in an:
aven / shaft   – if yes >> if no go to 3d</td><td colspan="2">What is the diameter of the tube (m):
Now go to Meso 4…</td></tr>
<tr><td colspan="2">Meso 3d – Is the roost feature in a:
hall / chamber / cave / cavern
    – if yes >> if no go to Meso 4</td><td colspan="2">What is the ceiling height (m):
Width on the north/south axis (m):
Width on the east/west axis (m):
Now go to Meso 4…</td></tr>
<tr><td colspan="3">Meso 4 – On the map you are using to navigate, does the meso-environment feature have a name?   – if yes >> if no go to Micro 1</td><td colspan="2">Write the name here:
Mark the roost location on the map</td></tr>
</table>

BASIC RECORD (cont.) / **WHAT 2.2**	
Micro 1 – Is the roost position in an exposed plane surface situation? yes / no – if yes complete the record below if no go to Micro 2a		
Bat hanging from: rock: flat surface, step or rib / brick: flat surface, step or rib / wood: crossbar – prop – cribbing / metal: beam, bearing plate or nail		
Bat species:		
Number of bats:	**Is the bat:** awake / torpid	
Height of the roost position above the floor (m):	**Droppings:** yes / no *Now go to 'associates'…*	
Micro 2a – Is the roost position or PRF in a simple recessed feature, such as a: pitch / cupola / manhole / vestibule / shelf / tent / ceiling pocket / wall pocket *– If the roost is in a double-recessed feature (e.g. a ceiling pocket in a cupola, or a wall pocket in a vestibule) the dimensions of the larger component are recorded below and those of the smaller component are recorded at 2b. If there is only an individual recess you may ignore 2b and go to 'associates'*		
Bat species:		
Number of bats:	**Is the bat:** awake / torpid	
Height of the roost position above the floor (m):	**PRF entrance aspect:** down-facing / outfacing / up-facing from floor	
Entrance dimension on longest axis (cm):	**Entrance dimension on shortest axis (cm):**	
Internal front-to-back width (cm):*	**Droppings below the bat:** yes / no *Now go to 'associates'…*	
Micro 2b – Is the roost position or PRF in a double-recessed feature such as a: ceiling pocket / wall pocket		
Entrance dimension on longest axis (cm):	**Entrance dimension on shortest axis (cm):**	
Internal front-to-back width (cm):*		
Micro 3 – List any associate invertebrate species present in the immediate vicinity:		

* Imagine you are looking at the dimensions of wardrobes at ikea.com – this is the depth measurement that would give the third dimension.

Table 5.3 The *BRHK Subterranean Rock Database* recording form – meso and micro values for enveloping roost positions.

WHO	1st Recorder:	2nd Recorder:	
WHERE	Site name: see Chapter 1		
WHEN	Date: see Chapter 1		

BASIC RECORD	**WHAT 2.1**	Meso 1a – Is the roost in the: threshold / dark-zone		Lux value:
		Meso 1b – What is the Distance between PRF and nearest entrance (m):		
		Meso 2 – Is the roost in the: ceiling / left wall / right wall / face / floor		
		Meso 3a – Is the roost feature: on a route / at a node or junction / in a destination *Now go to Meso 3b …*		
		Meso 3b – Is the roost feature on a route in a: portal / aisle / passage / keyhole passage / adit / tunnel / crawl / squeeze *– if yes >> if no go to 3c*	What is the ceiling height (m): What is the width between the walls (m): Is the roost feature: in a wider section of a route (i.e. you can walk past it) / at the end of a cul-de-sac / in a face *Now go to Meso 4 …*	
		Meso 3c – Is the roost feature in an: aven / shaft *– if yes >> if no go to 3d*	What is the diameter of the tube (m): *Now go to Meso 4 …*	
		Meso 3d – Is the roost feature in a: hall / chamber / cave / cavern *– if yes >> if no go to Meso 4*	What is the ceiling height (m): Width on the north/south axis (m): Width on the east/west axis (m): *Now go to Meso 4 …*	
	WHAT 2.3	Meso 4 – On the map you are using to navigate, does the meso-environment feature have a name? *– if yes >> if no go to Micro 1*	Write the name here: *Mark the roost location on the map*	
		Micro 1 – Is the roost feature sheltered within a simple recessed feature? yes / no *– If yes go to Micro 2, if no go to Micro 3 …*		
		Micro 2 – Circle the relevant recess type: pitch / cupola / manhole / vestibule / shelf / tent / ceiling pocket / wall pocket		
		Recess entrance aspect: down-facing / outfacing / up-facing from floor	Recess entrance dimension on longest axis (cm):	
		Recess entrance dimension on shortest axis (cm):	Recess front-to-back width (cm):*	
		Micro 3 – Which of the following is the bat occupying: a crevice / crack / break / fissure / pleat / alicorn / borehole / drain-hole / breakdown / choke / fill / other (please describe): *Now complete the record below …*		
		PRF height above or below floor (m):	PRF in: rock / brick / wood / metal	
		PRF entrance aspect: down-facing / up-facing / outfacing from a wall		
		What is the entrance dimension on the longest axis (cm):		
		What is the entrance dimension on the shortest axis (cm):		
		Bat species:	Number of bats:	
		Is the bat: awake / torpid	Are droppings present in the PRF: yes / no	
		Where is the bat in relation to the roost entrance: above / in front / to the side / below *If you are licensed to use an endoscope go to Micro 4 …*		
ADVANCED RECORD	**WHAT 2.4**	What is the distance between the entrance and the bat (cm):		
		What is the internal height (i.e. top of entrance to top of interior) (cm):		
		What is the internal width (i.e. inside entrance lip to the back wall in front*) (cm):		
		What is the internal depth (i.e. bottom of entrance to bottom of interior) (cm):		
		Is a comprehensive inspection possible: yes / no		
		Internal humidity: aridly dry / surface darkened by damp / obviously wet with runs or droplets		
		List any associate invertebrate species present inside:		

* Imagine you are looking at the dimensions of wardrobes at ikea.com – this is the depth measurement that would give the third dimension.

The values included on the first form comprise those needed for a basic landform record, and encompass the broad macro-environment variables. Unless there is a significant alteration to the situation this form will only need completing once.[31]

The second recording form is designed to anticipate exposed surface and recessed situations. The third form is designed to anticipate enveloped roost features.

Upon the first encounter of a bat on an exposed surface or in an open recess, it appears that the bat(s) have simply landed in the position by chance; that may be true, but this project is designed to investigate whether what we are seeing is in fact occupancy within the invisible niche.

The working theory is that the bat is in that spot because the air current there is exactly right, as is the temperature range and fluctuation, the humidity and the illumination. It may be that a move of even 10 cm above, below or on either side might put the bat in an unfavourable micro-environment. If this is accurate, a visit in the same week next year that is performed in broadly the same weather conditions as it was the year before will find that spot occupied by the same species of bat(s) that were recorded previously – no further in or out and no higher or lower.

This does not mean that specific positions will be occupied continuously: the evidence collated so far demonstrates that all species move in response to the advance of the seasons and the climate outside. The same bat might therefore be encountered in an exposed position in one week, in a recess the next week and in a completely enveloping PRF the week after that.

What it does mean is that those same roost positions will be occupied again next year in broadly the same order by the same species, and in all subsequent years and (if we record faithfully and the theory is not disproved) our descendants will be able to plan their visits to record the descendants of those bats, in exactly the same places.

The same will also be true of roosts recorded using the third form, which is more recognisable as a continuation of the rock face and loose rock recording forms because it anticipates the enveloping PRF where we have to peer into a narrow entrance, and perhaps even use an endoscope to complete the inspection. Notwithstanding, the top part of the form covers a good deal of meso-environment characteristics that will have to be very carefully recorded if the PRF is to be found again on another visit; it is easy to forget that the GPS upon which you rely in the wood to take you to the tree does not work underground.

In this project the decision was taken that the meso- and micro-environment record should amount to no more than one side of A4 and as much as possible it should be multiple choice for completion simply by circling.

Reference back to Chapters 3 and 4 will show that the subterranean recording form is more complex at the meso level but simpler at the micro-environment level. This is because the form is characterising what is essentially another world: at this stage of the project, we are focusing on the physical characteristics of that world, and the structure of the situations that bats occupy within it.

Because we can already anticipate that the greatest proportion of subterranean occupancy will be in the autumn-flux, winter and spring-flux (i.e. November through April) when bats are most vulnerable to disturbance, we want to keep physical intrusion to a minimum. When that is balanced against the need for detailed environmental data, it is clear that the deployment of automated temperature and humidity loggers will be more appropriate. This aspect of the micro-environment characterisation is discussed in detail in Chapter 6.

5.3.2 Introducing the subterranean bat species

Of the 18 species known to be resident in the British Isles, 17 are conclusively known to roost in subterranean landforms. At present there is insufficient data to produce anything meaningful (or even conclusive) in respect of Alcathoe's bat and this species is therefore the odd one out.

Another species, the greater mouse-eared bat, is excluded from detailed consideration in this chapter because there is only one greater mouse-eared bat known to occur in the British Isles. Devoting print space to a species for which there is little merit in creating habitat used a large amount of space that was better spent on other topics. This species was therefore omitted (at least in this edition).[32]

The remaining 16 species comprise:[33]

1. **Barbastelle** *Barbastella barbastellus* (Punt & van Nieuwenhoven 1957; Daan & Wichers 1968; Gaisler 1970; Mihál & Kaňuch 2006; Murariu & Gheorghiu 2010; Nagy & Postawa 2010; Piksa & Nowak 2013);

2. **Serotine** *Eptesicus serotinus* (Gaisler 1970; Daan 1973; Mihál & Kaňuch 2006; Nagy & Postawa 2010; Klys 2013; Piksa & Nowak 2013);

3. **Bechstein's bat** *Myotis bechsteinii* (Gaisler 1970; Mihál & Kaňuch 2006; Nagy & Postawa 2010; Piksa & Nowak 2013; Natural England n.d.);

4. **Brandt's bat** *Myotis brandtii* (Degn 1989; Gaisler & Chytil 2002; Mihál & Kaňuch 2006; Wermundsen & Siivonen 2010; Wermundsen 2010; Klys 2013; Piksa & Nowak 2013; Kristiansen 2018; Natural England n.d.);

5. **Daubenton's bat** *Myotis daubentonii* (Hooper & Hooper 1956; Punt & van Nieuwenhoven 1957; Bezem *et al.* 1960, 1964; Sluiter & van Heerdt 1964; Daan & Wichers 1968; Lesiński 1989; Gaisler 1970; Daan 1973; Degn 1989; Gaisler & Chytil 2002; Kokurewicz 2004; Mihál & Kaňuch 2006; Nagy & Postawa 2010; Wermundsen & Siivonen 2010; Wermundsen 2010; Klys 2013; Piksa & Nowak 2013; Kristiansen 2018; Natural England n.d.);

6. **Whiskered bat** *Myotis mystacinus* (Coward 1907; Pill 1951; Hooper & Hooper 1956; Punt & van Nieuwenhoven 1957; Bezem *et al.* 1960, 1964; Sluiter & van Heerdt 1964; Daan & Wichers 1968; Gaisler 1970; Daan 1973; Gaisler & Chytil 2002; Mihál & Kaňuch 2006; Wermundsen & Siivonen 2010; Wermundsen 2010; Klys 2013; Piksa & Nowak 2013; Kristiansen 2018; Natural England n.d.);

7. **Natterer's bat** *Myotis nattereri* (Hooper & Hooper 1956; Punt & van Nieuwenhoven 1957; Bezem *et al.* 1964; Sluiter & van Heerdt 1964; Daan & Wichers 1968; Gaisler 1970; Daan 1973; Gaisler & Chytil 2002; Mihál & Kaňuch 2006; Nagy & Postawa 2010; Wermundsen 2010; Klys 2013; Piksa & Nowak 2013; Natural England n.d.);

8. **Nathusius' pipistrelle** *Pipistrellus pipistrellus* (Altringham 2003; Dietz *et al.* 2011);

9. **Common pipistrelle** (Altringham 2003; Nagy & Postawa 2010; Murariu & Gheorghiu 2010; Dietz *et al.* 2011);

10. **Soprano pipistrelle** *Pipistrellus pygmaeus* (Altringham 2003; Murariu & Gheorghiu 2010; Dietz *et al.* 2011);

11. **Leisler's bat** *Nyctalus leisleri* (Schofield & Mitchell-Jones 2003);

12. **Noctule** *Nyctalus noctula* (Altringham 2003; Nagy & Postawa 2010; Murariu & Gheorghiu 2010; Dietz *et al.* 2011);

13. **Brown long-eared bat** *Plecotus auritus* (Hooper & Hooper 1956; Punt & van Nieuwenhoven 1957; Bezem *et al.* 1964; Gaisler 1970; Daan 1973; Yalden & Morris

1975; Gaisler & Chytil 2002; Altringham 2003; Mihál & Kaňuch 2006; Nagy & Postawa 2010; Wermundsen & Siivonen 2010; Wermundsen 2010; Klys 2013; Piksa & Nowak 2013; Kristiansen 2018; Natural England n.d.);

14. **Grey long-eared bat** *Plecotus austriacus* (Gaisler 1970; Gaisler & Chytil 2002; Mihál & Kaňuch 2006; Nagy & Postawa 2010; Klys 2013; Piksa & Nowak 2013);

15. **Lesser horseshoe bat** *Rhinolophus hipposideros* (Coward 1906, 1907; Hooper & Hooper 1956; Punt & van Nieuwenhoven 1957; Bezem *et al.* 1960, 1964; Sluiter & van Heerdt 1964; Daan & Wichers 1968; Gaisler 1970; Gaisler & Chytil 2002; Mihál & Kaňuch 2006; Schofield 2008; Nagy & Postawa 2010; Piksa & Nowak 2013; Natural England n.d.);

16. **Greater horseshoe bat** *Rhinolophus ferrumequinum* (Coward 1906, 1907; Hooper & Hooper 1956; Ransome 1968, Gaisler 1970; Ransome 1990; Mihál & Kaňuch 2006; Altringham 2003; Murariu & Gheorghiu 2010; Nagy & Postawa 2010; Natural England n.d.).

Long-term studies of ringed bats have proven that the species that occupy caves show long-term fidelity to a particular suite of sites that in combination offer them the environment they need over the course of the autumn, winter and spring months (e.g.

Table 5.4 The months in which individual bat species have been recorded occupying caves and mines (data and accounts taken from authors cited in the bulleted species list provided above).

SPECIES	WINTER		SPRING-FLUX	PREGNANCY			NURSERY		MATING		AUTUMN-FLUX	
	Jan	Feb	Mar	Apr	May	Jun	Jul	Aug	Sep	Oct	Nov	Dec
Barbastelle *Barbastella barbastellus*	✓	✓	✓	?	?	?	?	?	?	?	?	✓
Serotine *Eptesicus serotinus*	✓	✓	✓*	?	?	?	?	?	?	?	?	?
Brandt's bat *Myotis brandtii*	✓	✓	✓	✓	✓	?	?	?	✓	✓	✓	✓
Daubenton's bat *Myotis daubentonii*	✓	✓	✓	✓	✓	✓	?	✓	✓	✓	✓	✓
Whiskered bat *Myotis mystacinus*	✓	✓	✓	✓	✓	?	?	?	?	✓	✓	✓
Natterer's bat *Myotis nattereri*	✓	✓	✓	✓	?	?	?	?	?	?	?	✓
Brown long-eared bat *Plecotus auritus*	✓	✓	✓	✓	?	?	?	?	?	✓	✓	✓
Grey long-eared bat *Plecotus austriacus*	✓	✓	?	?	?	?	?	?	?	?	?	?
Lesser horseshoe bat *Rhinolophus hipposideros*	✓	✓	✓	✓	✓	✓	✓	✓	✓	✓	✓	✓
Greater horseshoe bat *Rhinolophus ferrumequinum*	✓	✓	✓	✓	✓	✓	✓	?	?	✓	✓	✓

* BRHK Subterranean Rock Database.

Bels 1952, Hooper & Hooper 1956, Daan & Wichers 1968). Bats are not, however, caverni-coles and therefore do not feature in texts dealing with cave biology. In his detailed accounts of subterranean biology, *Biospeleology*, Vandel (1965) has the following to say:

> *Many species of the genus Vespertilio and Plecotus hibernate in caves but are not found there during the summer. Rhinolophus are much more sensitive to low temperatures, and shelter during the summer and winter in caves. They are regular inhabitants of subterranean cavities.*

There is some Continental data for monthly occupancy and although other accounts do have wide gaps and there is much left vague (i.e. an irritating number of vague references to 'winter' but not specifically which month), by collating what is available a pattern can be seen in the encounters of some species and also shows where data are missing and are worth collecting. The result is provided in Table 5.4.

Precisely when Bechstein's bats, Nathusius' pipistrelles, common pipistrelles, soprano pipistrelles, Leisler's bats and noctules have been recorded in subterranean landforms is not published (i.e. we do not have specific month data). However, Nagy & Postawa (2010) reported that in Romania, barbastelles, serotines, Bechstein's bats, Daubenton's bats, whiskered bats, noctules and common pipistrelles were encountered in caves in the period December through March and also variously in May through August.

5.3.3 Introducing the subterranean rock 'What 1' macro-environment values

In the context of subterranean rock, the macro-level record comprises: **1)** the rock type; **2)** the landform itself; **3)** the topography the landform opens in from; **4)** the habitat over the opening through which the surveyors entered; **5)** the number of bat-friendly entrances the landform has; and, **6)** whether the landform holds any form of water feature.

5.3.4 Introducing the subterranean rock 'What 2.1' meso-environment values

The meso-environment is what a bat encounters as it flies through the landform and encompasses the multitude of different physical shapes and features inside that landform, and on which and in which will occur the environment in which it might roost.

The meso-environment inside the landform is divided into four categorical levels, as follows:

» **Meso 1** – We first consider the broad environment in terms of: **1a)** whether the roost position is within the threshold or the dark-zone; and, **1b)** the distance between the roost position occupied by the bat(s) and the entrance to the cave, mine or tunnel in metres from the nearest entrance;

» **Meso 2** – We record the broad orientation of the roost position in terms of whether it is on or in the ceiling facing down, on or in a wall or face and looking out, or in the floor and facing up;

» **Meso 3** – We then consider the broad situation the roost position is in;

» **Meso 4** – Whether the broad situation has a specific name on a map or plan.[34] This part of the record allows the roost position to be refound and reinspected.

5.3.5 Introducing the subterranean rock 'What 2.2, 2.3 and 2.4' micro-environment and roost account

While the meso-environment variables are common to both exposed and enveloped roost positions, the micro-environment record differs between the two. As a result, there are two different recording forms.

What 2.2: exposed roost positions and features

In the case of exposed roost features, the record anticipates that the bat(s) might be in: **a)** a simple exposed plane surface position (i.e. simply hanging from the ceiling or a wall); **b)** a simple recessed feature such as cupola or vestibule; or, **c)** a double-recessed feature such as a ceiling pocket within a cupola, or a wall pocket in a vestibule.[35]

If the bat is in an exposed plane surface situation, the record considers:

» **Micro 1 – a)** the substrate the bat is hanging on or from; **b)** the bat species; **c)** the number of bats; **d)** whether they are awake or torpid; **e)** the height the bat is above the floor; and, **f)** whether any droppings were present;

» **Micro 3** – Whether any associated invertebrate species were present.

If the bat is in a simple recessed feature, the record considers:

» **Micro 2a – a)** the form of the PRF the bat is occupying; **b)** the bat species; **c)** the number of bats; **d)** whether they are awake or torpid; **e)** the height of the roost position above the floor; **f)** the entrance aspect; **g)** the entrance dimension on longest axis; **h)** the entrance dimension on its shortest axis; **i)** the internal front-to-back width (if you were looking at the dimensions of an Ikea wardrobe this is the depth measurement that would give the third dimension); and, **j)** whether any droppings were present;

» **Micro 3** – Whether any associated invertebrate species were present.

If the bat is in a double-recessed feature the form and dimensions of the secondary component are also recorded, as follows:

» **Micro 2b – a)** Whether the secondary component is a ceiling pocket or wall pocket; **b)** the entrance dimension on longest axis; **c)** the entrance dimension on its shortest axis; and, **d)** the internal front-to-back width;

» **Micro 3** – Whether any associated invertebrate species were present.

What 2.3: enveloping roost features

The micro-environment recording criteria for the enveloping PRF comprises:

» **Micro 1** – Whether the PRF is within a recessed situation or opening into an exposed plane surface.

If the PRF is not within a recessed situation, Micro 3 is completed, but if it is in a recess then the following are recorded first:

» **Micro 2 – a)** The relevant recess type; **b)** the recess entrance aspect; **c)** the recess dimension on the longest axis; **d)** the recess dimension on the shortest axis; and, **e)** the recess front-to-back width.

If only a basic roost record is to be recorded, the values are as follows:

» **Micro 3 – a)** The form of the PRF; **b)** the height of the PRF entrance above the floor of the cave, mine or tunnel; **c)** the substrate the PRF is in; **d)** the aspect of the entrance; **e)** the entrance dimension on the longest axis; **f)** the entrance dimension on the shortest axis; **g)** the bat species that is occupying the PRF; **h)** the number of bats; **i)** whether the bat is awake or torpid; **j)** whether there are any droppings present; and, **k)** whether the bat is above the entrance (i.e. it has gone in and up), in front of the entrance (i.e. it has just gone straight in), to the side of the entrance (i.e. it has gone in and found a roost position slightly offset of the entrance), or below the entrance (i.e. it has gone in and down).

What 2.4: enveloping roost features – advanced record

The final part of this record might be considered the 'advanced' record because in the greater proportion of situations it will require the use of an endoscope. The advanced values comprise:

1. The distance between the roost entrance and the bat inside;
2. The internal height offered by the PRF;
3. The internal width offered by the PRF;
4. The internal depth offered by the PRF;
5. Whether a comprehensive inspection is possible;
6. A classification of the internal humidity measured in visually perceptible terms;
7. Whether there are any associated invertebrate species present in the PRF.

Note: There is no denying that this all makes for a complex record which takes time and often a good deal of physical effort to create. However, once a feature is mapped and recorded the information can be used to create a gazetteer to guide each future visit and this will change very little (if at all) for decades. The recording form itself can be created to anticipate the specific roost features in that landform, and might be simplified just to the Micro 1 bat record.

The following sections describe the individual values so the surveyor can recognise them and record them.

Note: Unlike Chapters 3 and 4, which had detailed data support, this chapter relies on general referencing. Unfortunately, the book was delayed, and delayed again, and then COVID-19 hit and meant that subterranean systems could not be visited due to the restriction on working in confined space, and the potential that the work might actually infect the bats. This situation has extended across 2020 and into 2021. As a result, while we do at least now have a recording framework we were not able to get the data we wanted prior to publication.

Rather than postpone publication again, the decision was taken to proceed with three sources of information, as follows: **1)** the small amount of data held on the *BRHK Subterranean Rock Database*; **2)** SSSI notifications in respect of: solution caves, sea caves, mines and railway tunnels; and, **3)** environmental data from studies in the British Isles and Europe.

5.4 What 1: recording the subterranean rock macro-environment values

5.4.1 The rock type

Thus far the limited data in respect of which rock types hold subterranean systems that are exploited by roosting bats are summarised in Table 5.5. N.B. Railway tunnels are the anomaly in this context and the rock type appears to have no influence on their occurrence or distribution. They are therefore not considered here.

Those people who are setting out to record the PRF in a specific face might identify the rock in advance using the British Geological Survey website.[36] The important thing to keep in mind is that not all solution caves are wholly limestone: some may encompass other rocks, such as the solution cave shown in Figure 5.10 which has a sandstone ceiling.

Figure 5.10 A solution cave that is limestone in the lower part of the system but has a sandstone ceiling. Top: looking along a short passage at the horizontal strata. Bottom: looking up into the ceiling which opens from limestone into sandstone ceiling pockets.

Table 5.5 The rocks types within which roosting bats have been recorded or reported occupying PRF, in subterranean situations.

ROCK TYPE	LANDFORM	RECORD HELD ON THE *BRHK ROCK FACE DATABASE*	RECORD IN WHITE PAPER OR PHOTOGRAPHIC ACCOUNT
Basalt	Solution cave	?	?
	Sea cave	?	?
	Mine	?	✓ – 1 x SSSI notification
Chalk	Solution cave	?	?
	Sea cave	?	?
	Mine	?	✓ – 9 x SSSI notifications
Granite	Solution cave	?	?
	Sea cave	?	?
	Mine	?	✓ – 1 x SSSI notification
Limestone	Solution cave	✓	✓ – 41 x SSSI notifications
	Sea cave	?	✓ – 7 x SSSI notifications
	Mine	✓	✓ – 28 x SSSI notifications
Mudstone	Solution cave	?	?
	Sea cave	?	?
	Mine	?	✓ – 6 x SSSI notifications
Quartzite	Solution cave	?	?
	Sea cave	?	?
	Mine	?	✓ – 1 x SSSI notification
Sandstone	Solution cave	✓	✓ – 3 x SSSI notifications
	Sea cave	?	?
	Mine	?	✓ – 9 x SSSI notification
Shale	Solution cave	?	?
	Sea cave	?	?
	Mine	?	✓ – 1 x SSSI notification
Siltstone	Solution cave	?	?
	Sea cave	?	?
	Mine	?	✓ – 2 x SSSI notifications
Slate	Solution cave	?	—
	Sea cave	?	—
	Mine	?	✓ – 3 x SSSI notifications
Tuff	Solution cave	?	?
	Sea cave	?	?
	Mine	?	✓ – 2 x SSSI notifications

5.4.2 The landform

The landforms have been described already and would not be difficult to separate. Table 5.6 provides a summary of the bat species that exploit each of the four. N.B. The table may not be complete, but it does provide a reliable starting point.

The recording form anticipates the four landform types so that the relevant type can simply be circled.

5.4.3 The entrance topography

The entrance topography values anticipate that the entrance into the landform might open in from: 1) a sheer face; 2) sloping ground; 3) level ground; 4) inside a dell or pit; or, 5) a cutting.

Faces opening in from sheer faces and sloping ground are typical of solution caves in the south-west. They are also typical of sea caves, mines and railway tunnels. Entrances dropping down from level ground are typical of solution caves in the north of England and also of ventilation shafts serving mines and railway tunnels. Whether the entrance

Table 5.6 Subterranean rock landforms that are proven to be exploited by individual bat species.

| BAT SPECIES | LANDFORM | | | |
| | NATURAL | | ARTIFICIAL | |
	Solution cave	Sea cave	Mine	Railway tunnel
Barbastelle *Barbastella barbastellus*	✓	?	✓	✓ (redundant and modified*)
Serotine *Eptesicus serotinus*	✓	✓	?	✓ (redundant and modified*)
Greater mouse-eared bat *Myotis myotis*	✓	?	✓	✓ (redundant and modified*)
Bechstein's bat *Myotis bechsteinii*	✓	?	✓	✓ (redundant and modified*)
Brandt's bat *Myotis brandtii*	✓	?	✓	✓ (redundant and modified*)
Daubenton's bat *Myotis daubentonii*	✓	?	✓	✓ (redundant and modified*)
Whiskered bat *Myotis mystacinus*	✓	?	✓	✓ (redundant and modified*)
Natterer's bat *Myotis nattereri*	✓	?	✓	✓ (redundant and modified*)
Common pipistrelle *Pipistrellus pipistrellus*	✓	?	✓	?
Brown long-eared bat *Plecotus auritus*	✓	?	✓	✓ (redundant and modified*)
Grey long-eared bat *Plecotus austriacus*	✓	?	✓	?
Lesser horseshoe bat *Rhinolophus hipposideros*	✓	✓	✓	✓ (redundant and modified*)
Greater horseshoe bat *Rhinolophus ferrumequinum*	✓	✓	✓	✓ (redundant and modified*)

* The tunnel has been modified to enhance it for roosting bats.

opens from a dell or pit is limited to solution caves and mines, but is of particular interest in this context because there is a perception that those that do open from dells or pits may be smaller and simpler but still offer comparable temperature stability as a much larger complex system with an exposed entrance; it appears that the pit offers a degree of buffering from air movement and this influences the meso-climate inside the system and thereon the microclimate at specific positions. Entrances accessed via cuttings are typical of mine adits and railway tunnels, but they do also occur where the seaward end of a sea cave roof has collapsed.

The recording form anticipates the various topographies so that the relevant one can simply be circled.

5.4.4 The habitat on/over the subterranean rock landform entrance(s)

The habitats on or over the entrance(s) are recorded using the Phase 1 criteria. The landforms themselves classify as follows:

» **Solution caves** are I1.5 – Rock exposure and waste / natural / cave;

» **Sea caves** are also I1.5 – Rock exposure and waste / natural / cave;

» **Mines** are I2.3 – Rock exposure and waste / artificial / mine;

» **Railway tunnels** are J5 – Miscellaneous / other habitat.

However, what is really wanted here is the habitat the landform opens out into and not the landform itself. In the case of sea caves this is likely to comprise 'H1 – Coastland / intertidal', with the appropriate tertiary classification. In all the other landforms there may be one habitat per entrance and if all the entrances are known, ideally the record will include them all and not just the habitat the entrance closest to the roost position.

> **Note:** Although it is labouring the point already made several times in the previous chapters, if we have the full 12-value alphanumeric OS grid reference, we can look at the habitat remotely using satellite images to see if there is anything common to roost situations occupied by specific bat species, and anything that separates them.

The recording form has only limited space and it may be more sensible to record the habitats on a separate sheet or by reference to satellite imagery so that the site visit can simply 'truth' the initial identification.

5.4.5 The number of 'bat-friendly' entrances

The number of entrances influences the environment inside the landform. In this context we are only interested in 'bat-friendly' entrances, i.e. those a bat can fly in through, so those that a bat can get in through but a person could not are included, but entrances that are totally flooded or entered through a waterfall etc. are not considered. Railway tunnel entrances should be mapped even if they are partially sealed: we are interested in airflow and bat flow.

In many cases the number of entrances and their locations can be identified using historical maps, some of which are available online. Almost all the solution caves in Scotland, Wales and England have been mapped, and the maps show the entrances and shafts, etc. These maps can be found and downloaded easily enough. Ireland appears to be well mapped, but the maps are less easily accessible. However, anyone considering performing a bat survey of a cave might contact the Speleological Union of Ireland who may be able to provide a map or a contact for an access request (see www.caving.ie).

Surprisingly, sea caves are the least mapped of all the subterranean habitats and anyone creating a map of a sea cave will be doing very useful work. Some mines are mapped, but the coverage is patchy and the maps themselves are typically incomplete and not to scale. The entrances to railway tunnels can be identified by reference to the Ordnance Survey sheet and online satellite imagery.

The number of entrances is simply noted on the recording form.

5.4.6 Water

Glover & Altringham (2008) found that of 53 solution caves, 32 (60%) were exploited by a suite of bats for swarming and they reported that the presence of flowing water was a significant negative factor. We want to know whether this holds true equally for roosting, and whether it is true for both flowing and still water, and all periods of the year.

We are therefore interested in whether the landforms in which bats are found roosting hold: **1)** free surface streams; **2)** sump streams; **3)** sumps; and, **4)** lakes.

- » **Free surface stream** – Any water flow over the floor of the system through which a surveyor might reasonably walk through in Wellington boots. Free surface streams are commonly encountered in adits which tend to have an upward incline into the mine specifically so that they will provide drainage. A photographic illustration of a free surface stream is provided in Figure 5.11.

- » **Sump stream** – Effectively a U-bend full of flowing water. Freezing-cold water flow can easily drag a surveyor off their feet and into a situation where they are taken under and drowned. As bats do not use scuba gear there is no reason to attempt to go through a sump stream.

- » **Sump** – A U-bend full of standing water. Often described as a water trap and deceptively dangerous in a mine context, a sump is a flooded throughway section where drainage water collects and is often too deep to enter without wading or swimming. Bats do roost over standing water, but even where the levels have dropped to a point where it appears possible to cross on foot BEWARE – silt gets trapped in sumps, and even where you can see the bottom assume you will be stepping into quicksand. A photographic illustration of three sumps that are roosts is provided in Figure 5.12.

- » **Lake** – A large subterranean water body that would only be sensibly traversed using a boat, typically within a cavern but sometimes spanning a hall in the form of a huge sump. Freezing cold and deadly dangerous. A photographic illustration of a lake is provided in Figure 5.13.

- » **No water** – Here we are talking about any situation that falls outside the four water categories listed. Puddles or damp silt, for instance, do not count as water features in the context of this attribute.

The recording form anticipates the four types so that where a water feature is present the relevant type may simply be circled.

5.5 What 2.1: recording the subterranean rock meso-environment values

5.5.1 Meso-environment 1a: roost situation and illuminance

Meso-environment 1a divides the landform into two zones, comprising: **1)** the threshold; and, **2)** the dark-zone. The threshold encompasses the part of the cave, mine or tunnel into which light penetrates sufficiently to allow the surveyor to move without the aid of

Figure 5.11 Free surface streams flowing through: top: a hall. Middle: an aisle. Bottom: a squeeze.

Figure 5.12 Top: a sloping silt-covered shelf on the left gives way to a deep and cold sump on the right; lesser horseshoe bats *Rhinolophus hipposideros* roost a little further on than the sump, which never floods to the roof of the passage. Left: clear water over a deceptively smooth silted base leads across a deep quicksand pool; greater horseshoe bats *Rhinolophus ferrumequinum* roost in the far end of this adit, which floods to about midway up the wall every winter. Right: the greater horseshoe bat *R. ferrumequinum* on the right wall is over a turbid silted floor that had to be checked very carefully with a pole. (Top image: Stuart Spray Wildlife)

Figure 5.13 Subterranean lakes in mine systems. To give an idea of scale, the opening on the far side of the top lake is an adit that a man can walk through, and the lower lake is 2 m below the shelf the man is standing on. Both lakes are over 50 m deep. (Top image: Stuart Spray Wildlife)

Figure 5.14 A greater horseshoe bat *Rhinolophus ferrumequinum* roosting in a sunlit position in the threshold of a solution cave.

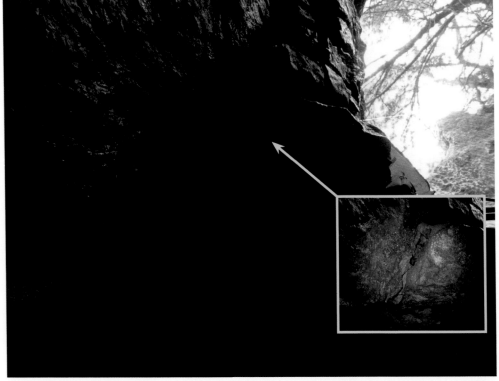

Figure 5.15 A lesser horseshoe bat *Rhinolophus hipposideros* roosting in a darkened position in the threshold of a solution cave.

a torch and any situation where a lux reading is above zero. The dark-zone is everything beyond the threshold and which is in permanent and complete darkness.

There is a common misconception that bats of all species roost in total darkness, but this is not true; even the horseshoe bats have no aversion to being in well-illuminated situations within the threshold, as with the greater horseshoe bat in Figure 5.14.

Other species will hide in plain sight in the dark-zone, by exploiting corners and projections, as in Figure 5.15.

Using the scientific accounts already listed in Section 5.3.2, an initial summary of roost situations has been created and provided in Table 5.7.

There are many factors that influence the extent of the threshold illuminance, such as: **a)** the topography the entrance opens in from; **b)** the vegetation the entrance opens in under; **c)** the height and width of the entrance; **d)** whether the entrance opens into a much larger void with a vaulted ceiling above; and, **e)** which direction the entrance faces.

Table 5.7 The bat species recorded and reported within the threshold and dark-zone offered within subterranean landforms.

SPECIES	THRESHOLD	DARK-ZONE
Barbastelle *Barbastella barbastellus*	✓	?
Serotine *Eptesicus serotinus*	✓*	✓
Bechstein's bat *Myotis bechsteinii*	✓	✓
Brandt's bat *Myotis brandtii*	?	✓
Daubenton's bat *Myotis daubentonii*	✓*	✓
Whiskered bat *Myotis mystacinus*	✓	✓
Natterer's bat *Myotis nattereri*	✓*	✓
Nathusius' pipistrelle *Pipistrellus nathusii*	?	?
Common pipistrelle *Pipistrellus pipistrellus*	✓	?
Soprano pipistrelle *Pipistrellus pygmaeus*	?	?
Leisler's bat *Nyctalus leisleri*	?	?
Noctule *Nyctalus nyctalus*	?	?
Brown long-eared bat *Plecotus auritus*	✓	✓
Grey long-eared bat *Plecotus austriacus*	✓	?
Lesser horseshoe bat *Rhinolophus hipposideros*	✓*	✓
Greater horseshoe bat *Rhinolophus ferrumequinum*	✓*	✓

* *BRHK Subterranean Rock Database.*

Not all these variables are recorded in the macro-environment record, because until we have established whether the environment value itself is important, we did not want to complicate the recording process any further than we already have. As a result, all we need is a lux value at the roost position measured using a light meter so that the tolerance of different bat species to light, can be identified.

5.5.2 Meso-environment 1b: roost distance from the entrance

Some published accounts do describe the distance from the entrance individual bat species have been recorded roosting from the landform entrance.[37]

As it can be predicted that the distance the bat will roost from the entrance will be determined by the climate inside and the location that offers a favourable roost environment, the primary reason the value is recorded is to make a mapping record for the specific landform. This will mean that each repeat survey is focused and dynamically looking for specific species in specific parts of the cave, rather than passively stumbling upon them.

5.5.3 Meso-environment 2: roost position

Meso-environment 2 considers whether the roost position is in: **a)** the ceiling; **b)** the right wall; **c)** the left wall; **d)** a face at the end of an adit, a capped shaft or a blocked tunnel, etc.; or, **e)** the floor.

In the greater proportion of situations this categorisation is simple, but it is sometimes difficult to decide where a sloping wall ends and an arched ceiling begins. This is because the walls may be reclined (i.e. leaning away from the surveyor), arched (i.e. leaning in towards the surveyor) concave or convex. It helps to look at the situation as though the ceiling were the floor and think where the wall would begin if you were Lionel Richie.[38]

The separation between a wall and a face is simple: a wall is parallel to a throughway or nodal feature such as a chamber, and a face is the perpendicular end. In solution caves faces typically comprise boulder chokes where the ceiling has collapsed or a node where the topography changes from a horizontal to a vertical throughway. Sea caves tend not to have faces because the passage decreases gradually in diameter into an impassable squeeze.

The term 'face' is taken from the mining context and it is here we find bare and planar surfaces. In mines a face rarely has any stone at the interface with the floor, but this is because it has all been removed; waste stone tends to be piled elsewhere. Photo 5.1 illustrates the clean working face at the end of an adit.

> **Note:** Some mining maps refer to the ceiling as the 'back' and the floor as the 'bottom'.

Obviously, railway tunnels only have a 'face' when an end has been bricked up.

The recording form anticipates the five situations in order that the relevant one may simply be circled.

5.5.4 Meso-environment 3: routes, junctions and destinations

Meso-environment 3 is where you are standing when you inspect the PRF to record the bat(s). The distribution of different species of bats occupying the same landform does not appear random. The perception is that those species that appear to favour exposed and recessed positions, congregate at specific points in specific systems. These appear superficially similar across multiple sites which suggests that the physical characteristics influence the environment favourably and would be worth attempting to replicate in an

Photo 5.1 The clean working face at the end of an adit; the walls are peppered with horizontal boreholes, as is the face itself.

artificial landform. The perception is also that specific enveloping PRF appear more common in some situations with the same general physical characteristics and are rare in others.

Meso-environment 3 categorises the roost situation into: **1)** a position that is on a route through the landform; **2)** a position that is on a junction or node that connects two or more routes; and, **3)** a position within a meso-feature that is a destination in itself.

A route is a throughway along which the survey will progress. A node or junction with be a functional fork in the throughway that allows the survey to progress in one direction or another. A destination is a feature beyond which the survey can proceed no further.

The division is decided by whether the survey will continue beyond the roost position, or whether it would turn back from this feature. All the physical features of the subterranean landforms hinge from this and there may be several attributes to a single roost position.

For instance, a bat hanging from the ceiling or on a wall at an indistinct point in an adit would be roosting on a route and other bats might fly past that point to roost deeper in the system. A bat roosting on the back wall of a T-junction might be ideally placed to be out of air movement, despite the fact that the route continued past it in three directions. Yet another bat might also roost in the same adit, but on the face at the end or in a small chamber, and therefore in a 'destination' because no other bat could fly further.

The recording form anticipates these three situations and allows the relevant one simply to be circled. After that, depending on the situation, the relevant values will be recorded in either: **Meso 4a** – Horizontal routes; **Meso 4b** – Vertical routes; or, **Meso 4c** – Destinations.

> **Note:** Wherever practical, the terminology adopted for each landform is that used by cavers, miners and railwaymen. It is the terminology that ecologists will encounter when consulting historical maps and written accounts.
>
> Cavers have a wider vocabulary of terms than miners, because most solution caves are more complex than typical mines. However, some mines have historically broken into solution caves, and some caving terms might be used to describe physical features in an atypical mine. These are at the discretion of the recorder, and will depend on the unique attributes of individual roost position in the specific landform they are attempting to describe.

> **Note:** Describing roost positions is a diagnosis and not an identification. The better the surveyor understands the processes by which the features are formed, the easier that diagnosis will be.

5.5.5 Meso 4a: horizontal routes

The first set of features we consider are the elongate types along which bats travel on a broadly horizontal plane. These comprise: **1)** portals; **2)** aisles; **3)** passages; **4)** keyhole passages; **5)** adits; **6)** tunnels; **7)** crawls; and, **8)** squeezes. These features are defined as:

» **Portal** – The outer wall of rock above and to the side of the mouth of a cave, mine or tunnel which is exposed to the elements and to daylight.

» **Aisle** – A throughway along which a surveyor can comfortably walk without the risk of banging their head, and in which the ceiling apex could not be physically reached without the use of a ladder or specialist equipment.

» **Passage** – A throughway along which it is possible to comfortably walk over most of its length with minimal crouching, although for much of the length it will be possible to touch the ceiling, or confidently inspect it with a torch and without the need for binoculars. Passages are typically wider than they are high.

» **Keyhole passage** – A throughway in a solution cave, the base of which has become incised. Keyholes typically have flowing water in the base, although sometimes this is limited to periods of heavy rainfall. Keyhole passages will often be traversed with feet on either side of a gully or deep fissure.

» **Adit** – A broadly level or slightly inclined mine passage that enters subterranean workings from the surface. Unlike a tunnel, an adit is always either blind, i.e. a cul-de-sac that finishes at a dead end, or connects with system of passages but offers the only entry/exit point into and out of the mine.

» **Tunnel** – A throughway that passes from the surface right the way through the subterranean landform, often from one side of a hill to another. Unlike an adit, a tunnel always has at least two open ends and allows entry by one portal and exit via another on one continuous route.

» **Crawl** – A low diameter throughway that must be negotiated on hands and knees.

» **Squeeze** – Any throughway that must be negotiated prone with the stomach and back in contact with the substrate.

Aisles and passages are easy to visualise, but crawls and squeezes less so; Figure 5.16 will remedy this and the images on the top illustrate that you can even get keyhole crawls.

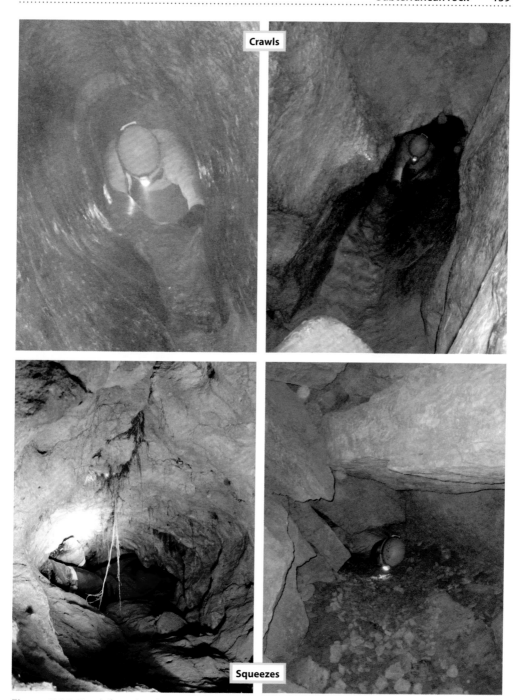

Figure 5.16 Top left and right: crawls that are becoming keyholed. Bottom left: a phreatic tube down which lesser horseshoe bats *Rhinolophus hipposideros* freely fly but tubby naturalists negotiate rather more slowly. Bottom right: a squeeze under a breakdown boulder.

The values recorded in these situations are measured at the position the bat is roosting in, and comprise: **1)** the ceiling height above the floor; and, **2)** the width between the parallel walls.

A bat may be roosting in one of these throughways but still be classed as being in a destination if it is within 2 m of the end of a dead-ended **cul-de-sac**, in a rock **face** at the very end, or in a **labyrinth** situation where a honeycombed network of throughways, **bridges** and **bridge-tubes** of varying diameters and lengths conglomerate. As a result, after the ceiling height and wall widths are recorded, a secondary categorical variable may be recorded, in a choice of: **a)** cul-de-sacs; and, **b)** labyrinths. These features are defined as:

» **Cul-de-sac** – Any cave throughway that doesn't go anywhere but culminates abruptly in an impassable obstacle.

» **Labyrinth** – A network of throughways and sometimes avens/shafts, etc. As labyrinths may comprise more than one horizontal level with wide open voids between sections, they may be functionally impossible to comprehensively access and inspect without installing permanent climbing anchors. Unlike solution cave labyrinths which tend to be uneven and comprised of columns,[39] pillars,[40] bridges[41] and bridge-tubes[42] as the result of lots of tributary flows entering the main-line passage, mines tend to have a series of level floors with passages and vertical shafts opening them up. Mine labyrinths are very dangerous because: **a)** the holes are big enough for a person to fall down them; **b)** the drop may be several metres deep; and, **c)** even short drops may end in water effectively leaving the surveyor in a well.

The recording form anticipates all the different features, allowing the relevant one(s) to be circled. However, additional values are also required, and in the elongate horizontal features the measurements taken are: **1)** the floor-to-ceiling height in metres; and, **2)** the wall-to-wall width in metres.

5.5.6 Meso 4b: vertical routes

Throughways also encompass vertical features. These comprise only two features: **1)** avens; and, **2)** shafts. These features are defined as:

» **Aven** – A medium-diameter vertical water-worn feeder tube extending upwards from the ceiling. Unlike shafts, although it may be possible for a surveyor to initially get up into an aven, the tubes constrict to the point a surveyor cannot access the full extent. Avens may be blind (i.e. dead-ended) or open to the world above. A photographic illustration of various avens is provided in Figure 5.17.

» **Shaft** – A high-diameter vertical tube which may drop vertically from the ground and by which the surveyor may enter the system below, or may extend downward from the floor inside the system. Unlike an aven, shafts remain wide enough for a surveyor to climb up and down inside and comprehensively explore with a torch. As with an aven, however, a shaft may be blind (i.e. dead-ended) or open. In the context of sea caves, a blow-hole is a shaft. A photographic illustration of two shafts is provided in Figure 5.18.

The recording form anticipates all the different features, allowing the relevant one to be circled. However, additional values are also required, and in the elongate vertical features, the measurement taken is the broad diameter of the tube in metres.

Figure 5.17 Left: looking up a blind aven. Right: looking up an open aven.

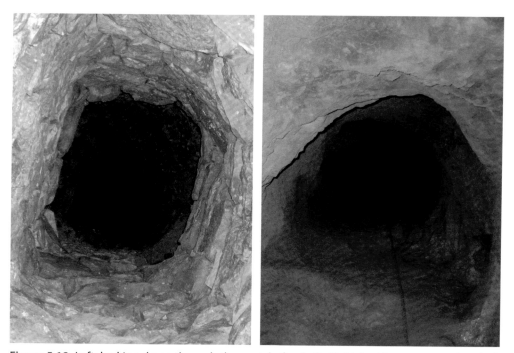

Figure 5.18 Left: looking down through the portal of a shaft. Right: looking up a complex shaft system from below.

5.5.7 Meso 4b: enlarged nodal and destination features

The enlarged nodal features are limited to solution caves, sea caves and mines. In solution caves, the nodal features are the places where water has pooled before percolating down, or pooled from water rising as ground water levels have risen and subsided. Sea caves tend to begin with a cavernous entrance which tapers into an aisle, which gradually tapers down to culminate in a squeeze. In mines the nodal features may be deep underground and accessed from adits. None of the railway tunnels visited by the project so far has had enlarged nodal features.

> **Note:** The enlarged nodal features are seductive in their distraction and that may prove deadly. All these voids immediately draw the eye to the ceiling but **STOP** at the entrance and look down: in that floor there may be a shaft somewhere...

The recording form anticipates four enlarged nodal features, comprising: **1)** halls; **2)** chambers; **3)** caves; and, **4)** caverns. These features are defined as:

» **Hall** – A hall is a rectangular cave- or cavern-like widened section on the line of a throughway such as an aisle or passage. Figure 5.19 illustrates halls in solution caves and mines.

» **Chamber** – A void that is wider than it is high and accessible to a surveyor, but has insufficient height for them to stand fully upright. Figure 5.20 illustrates a chamber.

» **Cave** – A typically room-sized void with a broadly flat floor, and with a ceiling that is sufficiently high for a surveyor to stand up comfortably. Unlike a cavern, other than where cryptic PRF are present and must be endoscoped, caves are sufficiently small that they be inspected with the use of a torch alone, and do not require binoculars or a telescope. Figure 5.21 illustrates solution caves.

» **Cavern** – A void that is larger than room-sized, and with the ceiling so high that it may be sufficient to cause a perceptible echo effect when talking. Unlike a cave, a cavern is so large that inspection of the ceiling and potentially the upper parts of the walls will require the use of binoculars or a telescope, and may be functionally impossible. Figures 5.22 and 5.23, as well as Photo 5.2, illustrate the difficulties faced in caverns.

The recording form anticipates all the different features allowing the relevant one to be circled. However, additional values are also required, and in the enlarged nodal features, the measurements taken are: **1)** the floor-to-ceiling height in metres; **2)** the width on the north/south axis in metres; and, **3)** the width on the east/west axis in metres.

Figure 5.19 Top: a typical solution cave hall. Bottom: the size that mining halls can reach (crying out for an evening of Edvard Grieg). (Bottom image: Stuart Spray Wildlife)

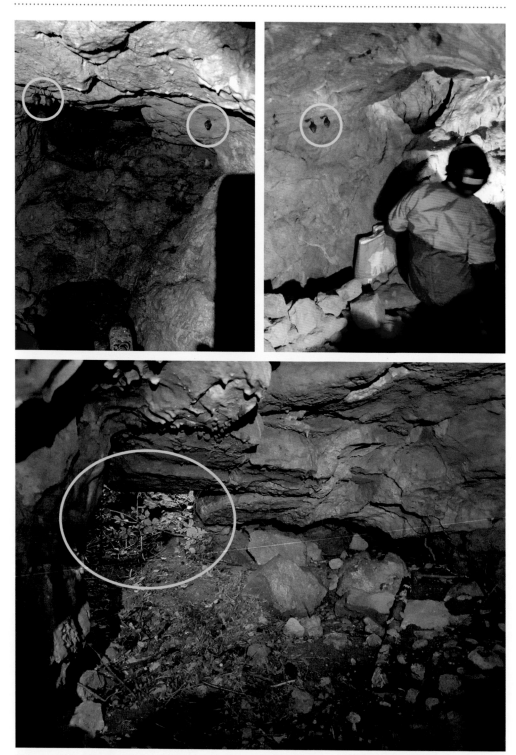

Figure 5.20 Chambers: the lesser horseshoe bat's *Rhinolophus hipposideros* domain. Top: do not be seduced by the obvious – while I was counting the lesser horseshoe bats in the chamber on the right, Louis Pearson found a Daubenton's bat *Myotis daubentonii* in a tiny crevice in the ceiling. Bottom: looking back at the entrance – this is the other side of the entrance shown in the bottom image in Figure 5.1.

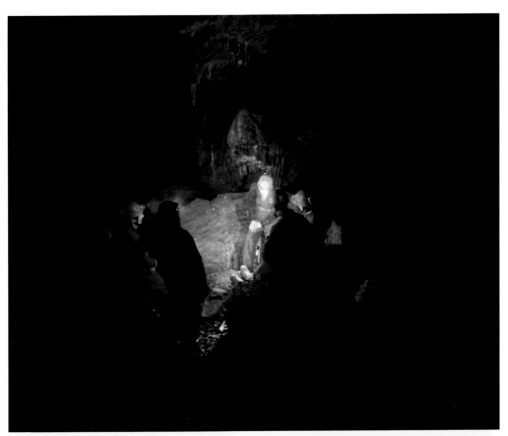

Figure 5.21 Top: A group of people having a Christmas party in a solution cave (you know who you are). Bottom: author's partner thinking, 'we had a day off from the children and you brought me down here so you could have … something to give some sense of scale!?!'

Figure 5.22 In some simple caverns the problem is one of getting the light in the places you need it. This cavern is big enough to drive a double-decker bus around in. (Both images: Stuart Spray Wildlife)

Figure 5.23 In more complex caverns the interface between the surface and subterranean environment is not clearly defined. (Both images: Stuart Spray Wildlife)

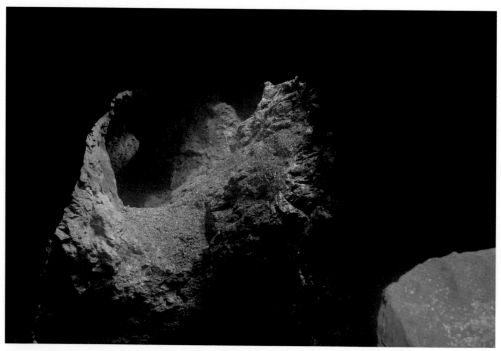

Photo 5.2 In some caverns the complexities are exacerbated by scale: the aisles in the upper part of the image are big enough to drive a bus through, but it would first have to jump over a pit 30 m wide sufficiently high to get up the 10 m shelf, and that gives a good idea of just how big the arch is in the bottom right of the image. (Stuart Spray Wildlife)

5.6 What 2.2: recording the subterranean rock micro-environment values for exposed plane surface and recessed roost positions

5.6.1 Micro 1: the bat roost record for exposed plane surface situations

The micro-environment record begins with the bat roost record for exposed plane surface situations. These comprise all situations where the bat is occupying an exposed situation that appears to have been selected entirely at random, inasmuch as there is an expanse of the same surface all around and nothing to distinguish the bat's position within it (viewed from a *Homo sapiens* perspective, it is not where we might choose to rest).

The record begins by identifying the substrate or particular structure from which the bat is hanging. These comprise: **1)** bare rock – flat surfaces, steps and ribs; **2)** brick lining; **3)** wood – crossbars, props and cribbing; and, **4)** metal – beams, bearing plates and nails.

Of these four substrates flat surfaces on bare rock and brick lining are easy to visualise, as I hope is a nail, but steps, ribs and the wood and metal mining features are less often encountered by bat surveyors and are therefore defined.

» **Steps** – Imagine you are under your stairs looking up: that is what a step in the wall of a mine or railway tunnel looks like. They are effectively a series of ceilings, but below the ceiling proper. Steps can be perpendicular to the wall, or parallel to it, as shown in Figure 5.24.

Steps are very much a feature of mines and railway tunnels where they can be large and chunky or narrow and shallow but they have no upper surface on which you might stand a cup of tea. Figure 5.25 illustrates narrow steps on the walls of mines.

Figure 5.24 Left: parallel steps. Right: perpendicular steps.

 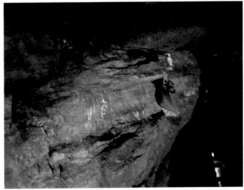

Figure 5.25 Left: a brown long-eared bat *Plecotus auritus* hanging from the underside of a step on the wall of a mine. Right: a greater horseshoe bat *Rhinolophus ferrumequinum* hanging from a step on the wall of a flooded adit.

» **Ribs** – A rib is an elongated horizontal outcrop extending from the wall, with a horizontal plane below and a horizontal plane above. Essentially it is like a fin extending along the wall. A rib can be just a seam of quartz extending along the wall of a mine adit, or a porch-canopy-sized projection where two phreatic tubes have merged to form a figure-of-eight-shaped passage in a solution cave, as in the example in Photo 5.3. There is potential for confusion between steps and ribs, so for the removal of doubt Figure 5.26 illustrates the difference.

» **Crossbar** – A horizontal wooden timber used to support the ceiling in a mine, typically held in place by upright timber posts (not a prop, which is different). Unhelpfully, a wooden support is referred to as a crossbar, but a steel support is referred to as a beam. N.B. A bat on a crossbar is in an exposed situation, unless it is between the crossbar and the rock in which case it is enveloped. Photo 5.4. illustrates a crossbar and post configuration.

» **Prop** – A single vertical timber used as a ceiling support in a mine in isolation and not as a part of a beam or crossbar configuration, which are held in place by posts. The project has never encountered a prop and had to resort to historical imagery for Figure 5.27.

Photo 5.3 Lesser horseshoe bat *Rhinolophus hipposideros* hanging from a rib.

Note: A bat roosting at the top of a flight of steps will be on the ceiling and not on the steps at all. And, a bat is only roosting on a rib if it is actually hanging from it. For the removal of doubt, Figure 5.26 shows the distinction between bats roosting on steps and ribs, from bats roosting on the ceiling.

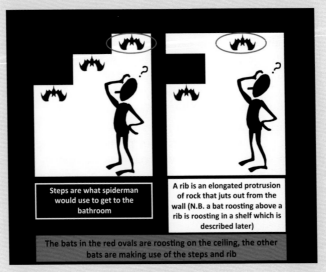

Figure 5.26 The distinction between bats roosting on steps and ribs, from those that are roosting on the ceiling.

Photo 5.4 A crossbar and post configuration. (© Yujnyj/Shutterstock)

Figure 5.27 The historical use of propping.

» **Cribbing** – A form of timber frame used to support the ceiling and also sometimes walls in a mine adit, etc. The most common form is the box-crib which is a tower structure, typically man-high. Photo 5.5 illustrates cribbing.

» **Beam** – A metal bar or girder used to support the ceiling of a mine. Beams may be fixed directly into the wall or held in place by upright posts (not a prop which is different). Unhelpfully, a steel support is referred to as a beam but a wooden ceiling support is referred to as a crossbar. N.B. A bat on a beam is in an exposed situation, unless it is between the beam and the rock in which case it is enveloped. Photo 5.6 illustrates the use of a beam.

» **Bearing plate** – A square of flat metal through which a $c.1$ m long roof-bolt is passed and driven into the rock to distribute the load and stabilise the **ceiling** or a **face** using steel mesh. The bearing plate itself is held on with a large nut. The entire combination of plate, bolt and nut are included under this one broad heading. N.B. A bat on a bearing plate is in an exposed situation, unless it is between the bearing plate and the rock in which case it is enveloped. Figure 5.28 illustrates the use of bearing plates.

The exposed plane surface roost record anticipates the different features allowing the relevant one to simply be circled. Thereafter the record encompasses: **1)** the bat species; **2)** the minimum number of bats visible; **3)** whether the bat is obviously awake or apparently torpid; **4)** the height of the roost position above the floor measured in metres; and, **5)** whether droppings are present below the bat. Of these values, whether there is any particular pattern in the height individual species roost above the floor has been looked at by several studies.

Pill (1951) reported that whiskered bats in Jug Holes Cave had the option to roost up to 3 m from the cave floor, but most chose positions that were near the floor and typically

Photo 5.5 A box crib. (© Mishainik/Shutterstock)

Photo 5.6 A beam.

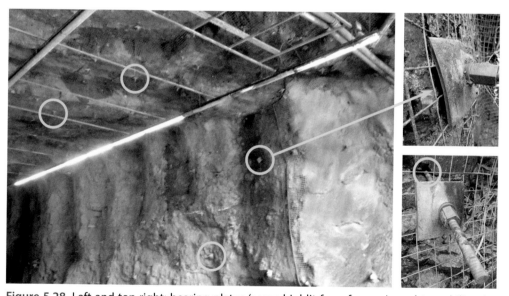

Figure 5.28 Left and top right: bearing plates (some highlit for reference) used to stabilise the walls and ceiling of a tunnel. Bottom right: where the bolt enters the wall it is in a borehole and this is a separate feature.

within 0.5 m of it. Cullingford (1962) summarises a great deal of work in the 1930s–1950s and notes that while lesser horseshoe bats favour low roost positions, greater horseshoe bats may roost as high as 15 m above subterranean floor. Bogdanowicz & Urbańczyk (1983) and Lesiński (1986) provide range data and Bezem *et al.* (1964) attempt to define the niche with a median across five years of detailed recording.

Table 5.8 Range of roost heights niches (m) above the floor occupied by the barbastelle, Serotine, Daubenton's bat, Natterer's bat, the brown long-eared bat and the grey long-eared bat.

SPECIES	Bogdanowicz & Urbańczyk 1983 (October – April) (m)	Lesiński 1986 (October – April) (m)
Barbastelle *Barbastella barbastellus*	0.4–6	1.5–5
Serotine *Eptesicus serotinus*	—	2.5–5
Daubenton's bat *Myotis daubentonii*	0–7.1	0.4–5.5
Natterer's bat *Myotis nattereri*	0.2–6.9	1.5–6
Brown long-eared bat *Plecotus auritus*	0.5–6	1.7–2.5
Grey long-eared bat *Plecotus austriacus*	—	1.5–2.4

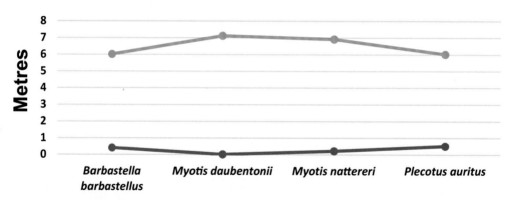

Figure 5.29 Minimum and maximum roost height range above the floor occupied by four species of bats (data taken from Bogdanowicz & Urbańczyk 1983).

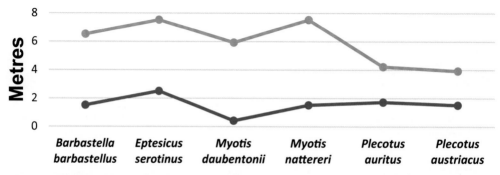

Figure 5.30 Minimum and maximum roost height range above the floor occupied by six species of bats (data taken from Lesiński 1986).

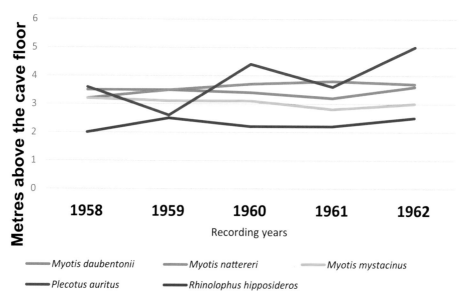

Figure 5.31 Median roost height above the floor occupied by five species of bats in a mine (data taken from Bezem *et al.* 1964).

The range data provided by Bogdanowicz & Urbańczyk (1983) and Lesiński (1986) are presented for use in Table 5.8 and illustrated in Figures 5.29 and 5.30. The niche data provided by Bezem *et al.* (1964) are provided in Figure 5.31.

Although the brown long-eared data were small and suggested fluctuations between years, Bezem *et al.* (1964) found that for all the other species the heights were stratified and stable within that stratification.

However, when they looked at the heights the bats might have occupied, they found that the lesser horseshoe bat only occupied roost situations where the ceiling was a median 2 m above the floor, the other four species were evenly distributed between median ceiling heights of 2, 3 and 6–7 m (Bezem *et al.* 1964).[43] What this means is that the tunnel ceiling might be high, but the bat species still occupied the same distance from the floor. The authors then looked at whether the height the bats chose to roost in was dictated by the position of specific exposed and enveloping roost features, and they found that regardless of what was on offer the stratification was constant and the niche partitioning in that specific mine was real, and held true across multiple years; the bats are selecting roost positions on the basis of something invisible, which is described in detail in Section 5.8.

5.6.2 Micro 2a: the bat roost record for recessed situations

The micro-environment values for recessed roost positions have to anticipate a simple situation where the bat is in an individual open feature, and also a slightly more complex situation where the bat is in a small feature that is itself within a larger one.

The features comprise: **1)** pitches; **2)** cupolas; **3)** manholes; **4)** vestibules; **5)** shelves; **6)** tents; **7)** ceiling pockets; and, **8)** wall pockets.

Of these ten features the two smallest – ceiling pockets and wall pockets – are those that may be encountered within larger recesses and thus offer double-recessed situations.

» **Pitch** – A peaked apex to the ceiling of an aisle, passage or cave, etc. Effectively the shape of the roof of a house as seen from the loft looking up. They may be acute and high or obtuse and short. Pitches are a particularly common feature of sea caves

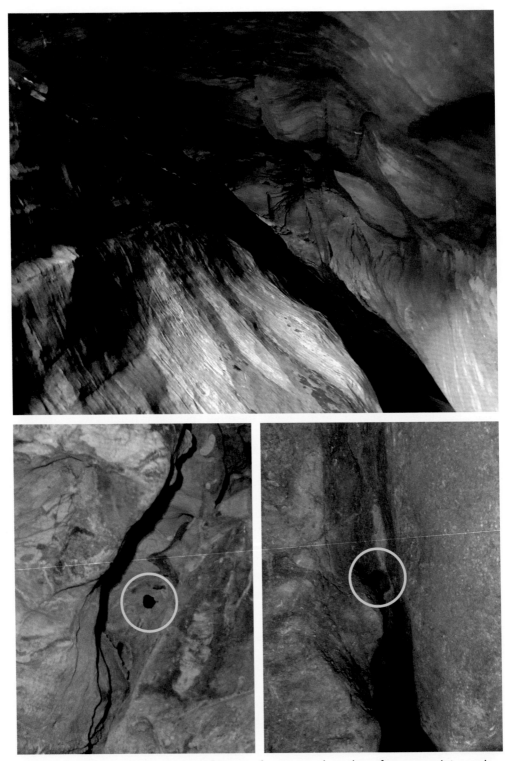

Figure 5.32 Top: pitches are common features of sea caves where they often narrow into crevices which bend off to one side; very difficult to inspect even where the ceiling itself is assessable. Bottom: pitches in solution caves are typically straightforward to survey with a torch; these have greater horseshoe bats *Rhinolophus ferrumequinum* roosting in them.

where they may extend up into a crevice. Figure 5.32 illustrates pitches in sea and solution caves.

» **Cupola** – Either a large-diameter (i.e. basket-ball size and above) rounded solution-dome in the ceiling or a wide and elongate almond-shaped recess. Photo 5.7 illustrates a cupola.

Some elongated cupola are difficult to separate from pitches. The division is best made by thinking in terms of air movement. In a pitch, the air will be channelled along the length and the bat might actually be in a concentrated draught of air. In a cupola, the air movement would pass across the ceiling and across the entrance to the feature, but the bat would be above the draught of air. The two situations are illustrated in Figure 5.33.

Photo 5.7 A cupola with a lesser horseshoe bat *Rhinolophus hipposideros* in residence.

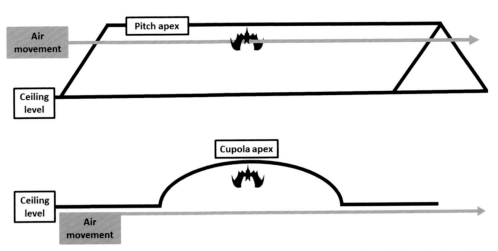

Figure 5.33 Separating pitches from elongate almond-shaped cupola by whether the bat is in a draught or above the draught.

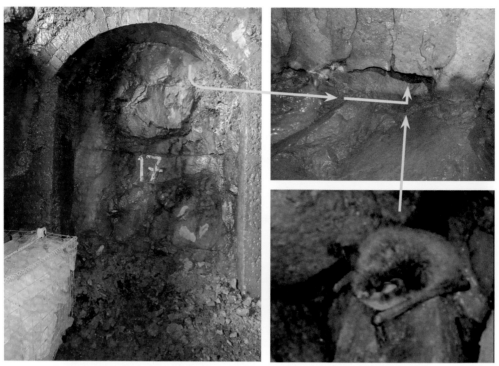

Figure 5.34 A crevice in the back of a manhole that is exploited by a Daubenton's bat *Myotis daubentonii*.

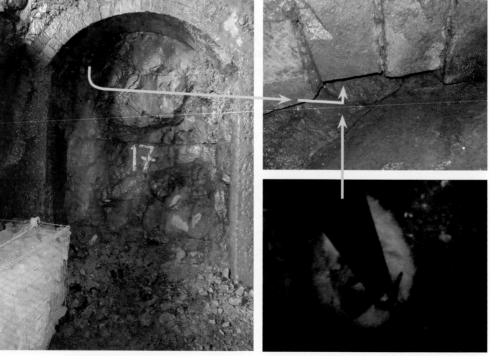

Figure 5.35 A crevice in the back of a manhole that is exploited by a Natterer's bat *Myotis nattereri*.

» **Manhole** – The mining and railway tunnel equivalent of a vestibule. A shallow but tall recess in the wall of a mine adit or railway tunnel in which workers can stand aside of a route. A refuge manhole might be thought of as a large alcove rather like a sentry-box. In mines, manholes allow a miner to get out of the way of a load of ore being transported to the surface (typically in a small man-powered truck on rails). In railway tunnels they were required to give safe shelter to groups of workers from passing locomotives, and a single manhole may accommodate two people or more. In mines they are just bare rock, but in railway tunnels the sides are typically brick-lined, and where the mortar is gone the gaps may give a bat access into crevices and cracks in the rock behind. Figures 5.34 and 5.35 illustrate a refuge manhole that holds crevices exploited by both Daubenton's and Natterer's bats.

» **Vestibule** – A natural shallow but tall recess in the wall of a solution cave that is big enough for someone to stand in it. A vestibule might be thought of as a large alcove rather like a natural manhole. Unlike a chamber or tent, there is room to stand up in a vestibule. N.B. The term was coined by Balch (1937). Figure 5.36 illustrates a vestibule.

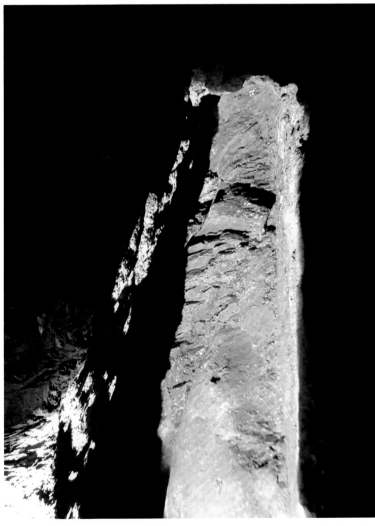

Figure 5.36
A vestibule.

» **Shelves** – A shelf is the direct opposite of a rib and constitutes a horizontal recess extending into the wall. Just as with ribs, shelves can vary in scale and may be encountered as narrow grooves and spaces long enough for a surveyor to lie in.

The division between a rib, and a shelf is not always easy to see. In particular, confusion may come about as a result of the fact that a rib may have a pronounced shelf above, with the upper surface of the rib forming the base of the shelf. Figures 5.37 and 5.38 illustrate the difference between a rib and a shelf and how a rib never exists in isolation but always has a shelf above.

A bat roosting above a rib, i.e. between the rib and the ceiling, is roosting in a shelf.

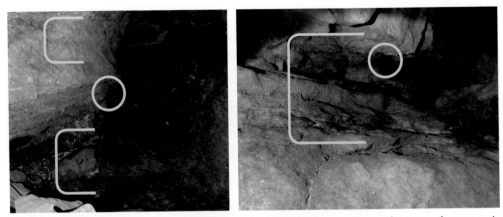

Figure 5.37 Left: a rib results in a shelf above, but this is often indistinct because the recess is rounded. Right: a shelf will not have a completely level base that you could stand a ball on without it rolling off, but it is clearly cut into the wall and much more pronounced recess than that above a rib. Both are occupied by lesser horseshoe bats *Rhinolophus hipposideros*.

Figure 5.38 The difference between a rib and a shelf.

Photo 5.8 A tent going off from the floor of a cave. Tents appear to originate from historical tributary inflow. This example is large and was chosen because the photograph came out well, but they may be much narrower and constricting: in one notorious feature a surveyor typically pushes themselves in by a second surveyor sitting down to give them a substrate to push on. When the inspection is finished the second pulls the first out by grabbing their ankles and heaving.

» **Tent** – A small triangular short section crawl that abruptly decreases in width as the floor rises to the apex. The feature is so called because it typically has a broadly flat floor and a pitched apex. Unlike a vestibule, a tent is longer than it is high and offers insufficient room for a surveyor to stand. The feature appears to originate from tributary inflow and then silting and might be thought of as horizontal aven. They are seldom more than 3 m in length. The feature is popular with lesser horseshoe bats and as the ceiling is often ruffled it is easy for a surveyor to disturb the bats as they enter the tent to search it. It is therefore best for the surveyor to enter the feature on their back and head first: uncomfortable for the surveyor but a lot less uncomfortable for the bat! Photo 5.8 illustrates a tent extending into the wall from the floor of a cave.

» **Ceiling pocket** – Any recess in the ceiling. In a solution cave or a sea cave a ceiling pocket will typically be rounded or almond-shaped and smaller than a football or rugby ball. Ceiling pockets may occur in isolation, or in a pitch, or in a much larger cupola. In a mine, ceiling pockets range from cube/rhomboid, jagged pyramidal and drilled cylinders (the latter a feature of adit ceilings). In all the examples of railway tunnel ceiling pockets recorded by the project, they have been the result of individual bricks being omitted to allow drainage through from above. Figure 5.39 illustrates the smooth ceiling pockets in solution caves and Figure 5.40 illustrates the rhomboid and cylindrical ceiling pockets in mines.

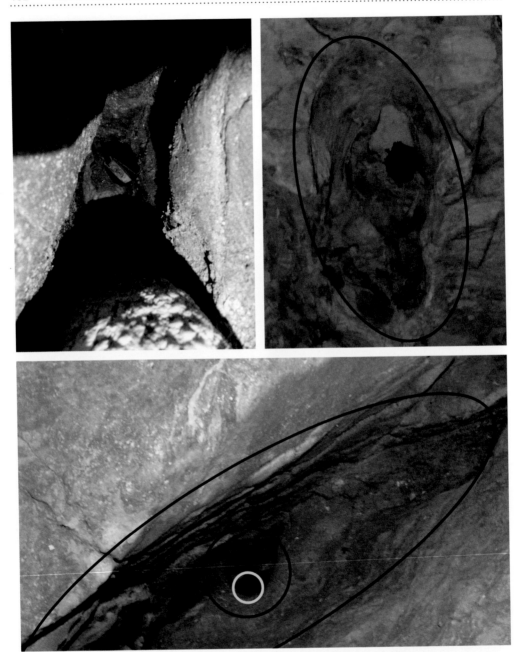

Figure 5.39 Top: smooth ceiling pockets in solution caves. Bottom: a ceiling pocket within an almond-shaped cupola. All are occupied by greater horseshoe bats *Rhinolophus ferrumequinum*.

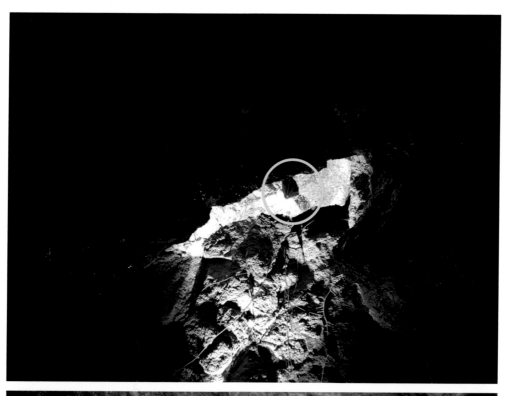

Figure 5.40 Top: a rhomboid ceiling pocket in the threshold of a mine occupied by a lesser horseshoe bat *Rhinolophus hipposideros*. Bottom: a cylindrical ceiling pocket in an adit.

» **Wall pocket** – Any recess in a wall that is large enough that a surveyor could put their fist in it, but smaller that a football in diameter. They are easily identified and inspected in caves and mines, but often overlooked in railway tunnels, where they comprise missing bricks in the lining and where they may hide bats in recesses off to one side of the entrance. Photo 5.9 illustrates a typical wall pocket in a solution cave.

In mines a wall pocket can be easily missed as just a dent in the surface, but some of those dents are used for sufficient periods to be identifiable when the bats are absent, even when occupation is only one bat! Figure 5.41 illustrates a wall pocket occupied by an individual whiskered bat, the black label-line shows where the bats use of the feature has resulted in the landing surface becoming scratched leaving a dull matt surface among the otherwise shiny surface that surrounds it.

The simple recessed roost record anticipates the different features allowing the relevant one to simply be circled. Thereafter the record encompasses: **1)** the bat species; **2)** the minimum number of bats visible; **3)** whether the bat is obviously awake or apparently torpid; **4)** the height of the roost position above the floor measured in metres; **5)** the recess entrance aspect (e.g. in the ceiling opening downward, or in a wall opening outward, etc.); **6)** the entrance dimension on the longest axis in centimetres (e.g. a vestibule and manhole this would be on the vertical axis); **7)** the entrance dimension on the shortest axis in centimetres (e.g. on the vestibule and manhole this would be on the horizontal axis); **8)** the internal front-to-back width in centimetres (it may help to imagine you are looking at the dimensions of wardrobes at ikea.com – this is the depth measurement

Photo 5.9 A typical wall pocket in a solution cave occupied by a lesser horseshoe bat *Rhinolophus hipposideros*.

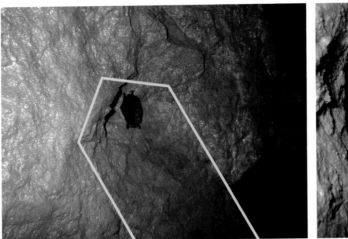

Figure 5.41 A small Myotis bat (most likely whiskered bat *Myotis mystacinus*) occupying a wall pocket in a mine. The image on the left has been annotated to show the landing surface, which is conspicuously dull and matt in contrast with the shiny substrate all around it.

for the third dimension and how far the wardrobe will stick out from the wall against which it will stand); and, **9)** whether droppings are present below the bat.

Obviously if the entrance to the feature is a perfect circle then the entrance dimension will be the same value (i.e. a diameter value).

5.6.3 Micro 2b: the bat roost record for double-recessed situations

A double-recessed situation would still first require the recording of all the values at Micro 2a, then the values specific to the secondary smaller feature the bat is actually inside. These comprise: **1)** the feature itself which may simply be circled or written in the cell; **2)** the entrance dimension of the smaller feature across the longest axis in centimetres; **3)** the entrance dimension of the smaller feature across the shortest axis in centimetres; and, **4)** the front-to-back width in centimetres.

5.6.4 Micro 3: the associate species

The Micro 3 value relates to any invertebrate species that might be near to the bat and experiencing the same micro-environment.

The species of bats that exploit caves can be described as 'trogloxenes' inasmuch as they visit caves seasonally, but do not complete their whole life cycle there (Chapman 1993). Even the bat species that may be encountered roosting in a subterranean situation year-round forage above ground.

There are, however, 'troglophile' invertebrates that can complete their life cycle in caves and also in non-cave habitats, and 'troglobites' that are obligate cave-dwellers and cannot survive outside caves (Chapman 1993). The cave spider *Meta menardi* is an example of a troglophile (see Figure 5.42).

The optimum temperature for the common cave spider *Meta menardi* is 6–8 °C (Vandel 1965). The species is also sensitive to drying and occupies situations of high humidity (Vandel 1965). Sparrow (2009) describes the discovery of a Mendip cave as the result of a caver noticing a cave spider egg-case beneath a small overhang on the surface, and the known temperature and humidity niche offer a useful bench-mark for predicting the presence of roosting bats.

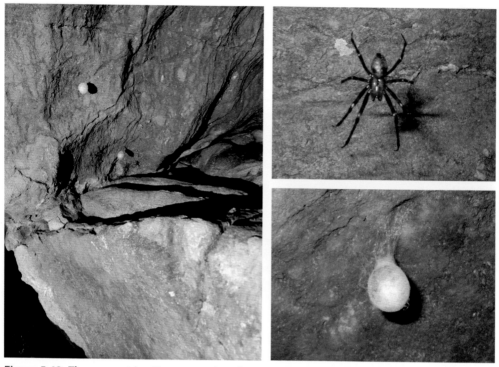

Figure 5.42 The cave spider *Meta menardi* and its conspicuous egg-cases.

The spider is present year-round, is conspicuous and has been recorded in all four of the subterranean landforms, even open railway tunnels where it occupies damp manholes. The spider is therefore useful as an environment 'surrogate' for the bats themselves.

What we are getting at here is that if the spider is present, then that is a good sign and increases the optimism that bats may be present.

Two species of moth are also cited as indicators of cave presence: the herald *Scoliopteryx libatrix* and tissue *Triphosa dubitata* (Sparrow 2009). Figure 5.43 shows what to look for.

At the moment we do not know what herald and tissue moths favour in terms of air movement, temperature and humidity, but if we record this we soon will and can then compare the data with the environmental niche occupied by specific bat species to see whether there is any useful association.

If associations between specific invertebrate troglophile and troglobite species were identified, the invertebrates might serve as a surrogate for the bats and demonstrate that conditions were suitable within the cave, even in the absence of any physical records of the bats themselves.

Finally, although individual bat species occupy discrete environmental niches, some of those niches have overlaps and these might be used to infer suitable conditions across species. For example, Bogdanowicz (1983) reported statistically significant associations between the barbastelle and brown long-eared bat, and between Daubenton's bat and Natterer's bat. In both cases, the paired species were found in mixed species clusters.

Figure 5.43 Left: the herald *Scoliopteryx libatrix*. Right: the tissue *Triphosa dubitata*. (Right image: Paul Bowyer – Somerset Moth Group)

5.7 What 2.3: recording the subterranean rock micro-environment values for enveloping roost positions

5.7.1 Micro-environment 1: enveloping PRF in plane surface situations

The first distinction to make when recording enveloping roost features is whether they are themselves in a recessed feature or an exposed situation going into a broadly planar surface such as a long or wide ceiling, high, long or wide wall or the floor. This is a simple dichotomy: if the roost feature is itself within a recess the recess is recorded before the enveloping PRF.

5.7.2 Micro-environment 2: enveloping PRF inside recesses

A thorough inspection will reward with a significant number of enveloping roost features. However, a thorough inspection will require the surveyor to get into the recess and inspect it from different physical aspects, such as with their back to the back wall facing out of the recess, crouching down and even lying on their back in order to look up. A useful trick is to imagine your hand is the bat, place your hand on the rock surface where a bat might land and then crawl your hand around to find entrances that might be hidden in little shadows left in the beams of torches and headlights: when searching for enveloping PRF, eyes are often subordinate to touch.

The recessed features have already been described at Section 5.6.2 and comprise: **1)** pitches; **2)** cupolas; **3)** manholes; **4)** vestibules; **5)** steps; **6)** ribs; **7)** shelves; **8)** tents; **9)** ceiling pockets; and, **10)** wall pockets.

In the context of an enveloping roost feature that is within a recess, the record requires the following: **1)** the recess entrance aspect (e.g. in the ceiling opening downward, or in a wall opening outward, etc.); **2)** the entrance dimension on the longest axis in centimetres;

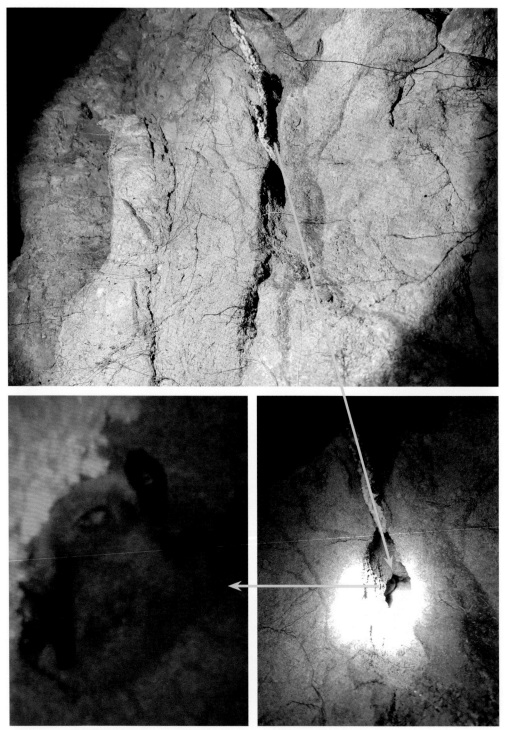

Figure 5.44 A Daubenton's bat *Myotis daubentonii* recorded roosting in a crevice in a solution cave ceiling by Louis Pearson.

3) the entrance dimension on the shortest axis in centimetres; and, **4)** the internal front-to-back width in centimetres.

5.7.3 Micro-environment 3: the enveloping PRF itself

The enveloping PRF comprise: **1)** crevices; **2)** cracks; **3)** breaks; **4)** fissures; **5)** pleats; **6)** alicorns; **7)** boreholes; **8)** drain-holes; **9)** breakdown; **10)** choke; and, **11)** fill.

> » **Crevice** – Any narrow and elongated cutting extending upward into the ceiling. Figure 5.44 illustrates a crevice in a solution cave ceiling and Photo 5.10 illustrates a typical crevice in the pitched ceiling of a sea cave.
> » **Crack** – A narrow vertical cutting that extends perpendicular into a wall or face. Photo 5.11 illustrates a typical crevice roost in a solution cave (see Chapter 3 for other photographic examples of crack roost features).
> » **Break** – A horizontal cutting that extends perpendicular into a wall or face. Photo 5.12 illustrates a typical break roost in a solution cave (see Chapter 3 for other photographic examples of break roost features).
> » **Fissure** – A narrow and elongated cutting extending down into the floor in the same way a crevice extends up into the ceiling (apologies for the lack of photos).
> » **Pleat** – A helictite comprising a ruffled curtain of limestone, often incorporating lots of stalactites. Figure 5.45 illustrates complex and simple pleats. Figure 5.46 and Photo 5.13 illustrate how Vespertilionid bat species use pleats.

Photo 5.10 A crevice in the apex of a pitched sea cave ceiling; these are often broken into a series of narrow pockets that may be variously discrete or connecting.

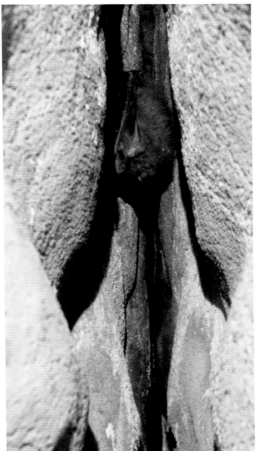

Photo 5.11
A brown long-eared bat *Plecotus auritus* in a crack in a solution cave wall. (Rich Flight – Flight Ecology)

Photo 5.12 Three Daubenton's bats *Myotis daubentonii* in a break in a solution cave wall. (Rich Flight – Flight Ecology)

Figure 5.45 Top: complex curtains of pleats require an endoscope to search. Bottom: simple pleats may be searched with a torch. Both are occupied by lesser horseshoe bats *Rhinolophus hipposideros*.

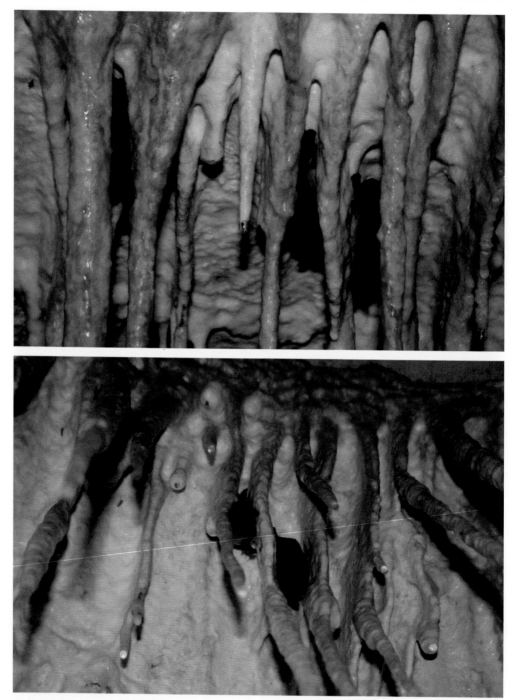

Figure 5.46 Complex, fragile and wet pleats occupied by whiskered bats *Myotis mystacinus*. (Both images: Colin Morris)

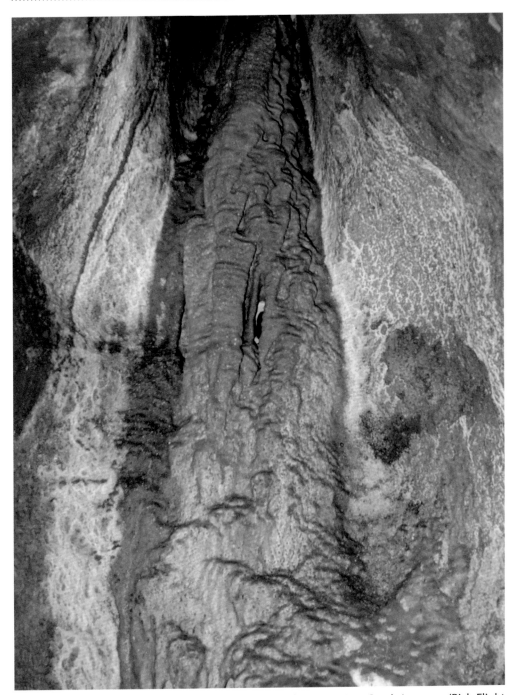

Photo 5.13 An unidentified bat roosting in a pleat in flowstone in a Cumbrian cave. (Rich Flight – Flight Ecology)

» **Alicorn** – A tall, narrow and sinuous solution tube extending up into the cave roof and typically just wide enough for an individual Vespertilionid bat to roost in the apex. An alicorn is best thought of as a miniature aven. Figure 5.47 shows how near to an aven they are; the difference is that an alicorn is too narrow to even post a small child inside (even if his mother was as committed to unravelling the mysteries of roosting ecology as his father is, and would allow him to be posted up there in the first place).

In most cases, however, alicorns are much smaller in diameter. Figure 5.48 illustrates alicorns that could be inspected with a torch, and Figure 5.49 illustrates a typical palm-sized alicorn to show the characteristic oval roost entrance, and what you may find if you have an endoscope to complete the inspection. Figure 5.50 demonstrates how beyond a typically oval entrance, there can be many weird and wonderful configurations that you will only see if you take an endoscope into the depths.

» **Borehole** – A deep hole that has been drilled into the rock in a mine. Although some boreholes are in the ceiling and made to take the bolts of bearing plates, most were bored to break out the rock, and are visible on the direction of travel but almost invisible when returning the other way. The feature typically comprises a cylindrical cutting and a relatively shallow hole, typically under 10 cm deep. Figure 5.51 illustrates boreholes in the walls and ceiling of an adit.

They can look really mucky on the outside, but actually be surprisingly clean within, like a horizontal alicorn, as in the one shown in Figure 5.52.

» **Drain-holes** – These are openings built into the ceiling and walls of railway tunnels to direct seasonal water flows. The holes were built in at regular intervals and may not be intended to direct any specific flow. Figure 5.53 shows typically wet drain-holes, but these are actually the exception rather than the rule.

In dry drain-holes the bats are immediately obvious upon inspection with a torch, such as the example in Figure 5.54.

At first glance, drain-holes may appear to be effectively ceiling pockets and wall pockets in the brickwork, but all is not always as it seems. Drain-holes often take the flow into the pocket from the sides, and have enveloping features off to one side of the pocket. What appears to be empty may hold a bat in a compact feature off to one side of the back wall of the pocket, as in the serotine roosting in the example in Figure 5.55.

» **Breakdown** – Boulders that have fallen from the cave or mine ceiling, which is ultimately left domed as a result. Breakdown is effectively subterranean talus but will not entirely block the cave; this would be a choke. Photo 5.14 illustrates breakdown.

» **Choke** – Any situation where the system is constricted or terminated by fallen rock. This rock itself may act as a PRF and, providing the situation is safe both for the surveyor and the bats, might be searched without disturbing the boulders by using a torch and endoscope. The feature is identical in appearance to breakdown, only tends to have larger boulders in with the cobbles and takes the form of a talus cone, stretching from the floor to the ceiling. Photo 5.15 shows an active choke.

» **Fill** – Unlike breakdown and choke, which are the result of a collapse, fill encompasses waste stone used to support the ceiling of a mine following the removal of saleable material, and just dumped muck. It is easy to just shine a torch down an adit, see fill and not bother checking the ceiling above it. This is a rookie mistake! It is surprising how many bats can be in a ceiling pocket above the sludge, or just on a wall and simply in a shadow. Figure 5.56 illustrates fill.

Figure 5.47 An alicorn with a lesser horseshoe bat *Rhinolophus hipposideros* roosting in the base: the bat is more often about halfway up and just visible behind a rib in the limestone. The inset image shows the feature within the apex of a larger cupola.

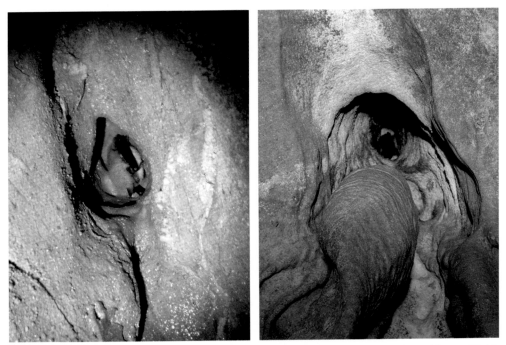

Figure 5.48 Left: an open alicorn roost with a Natterer's bat *Myotis nattereri* in residence. Right: another open alicorn with an unidentified *Myotis* sp. in residence. (Both images: Rich Flight – Flight Ecology)

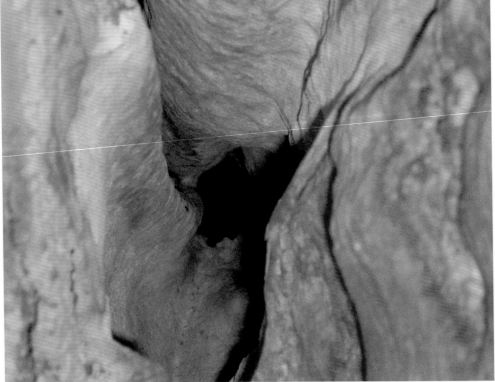

Figure 5.49 An alicorn roost with a Natterer's bat *Myotis nattereri* in residence.

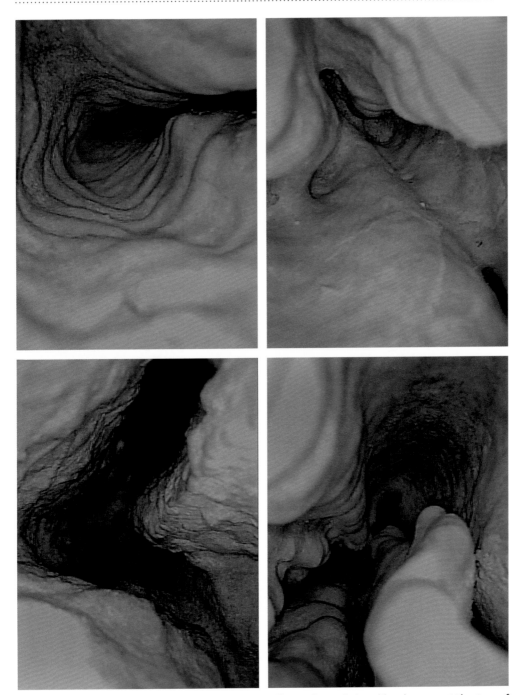

Figure 5.50 The entrances to these alicorns are all simple ovals just like the one at the top of Figure 5.49, but once inside there are a multitude of different shapes to investigate.

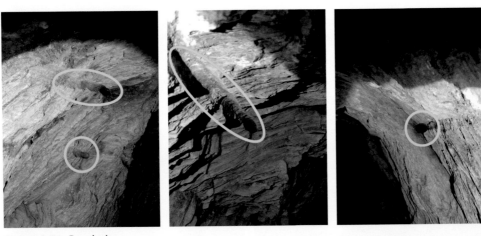

Figure 5.51 Boreholes

Figure 5.52 Left: view of a typical borehole outside. Right: the clean and oddly polished interior of the borehole seen though an endoscope.

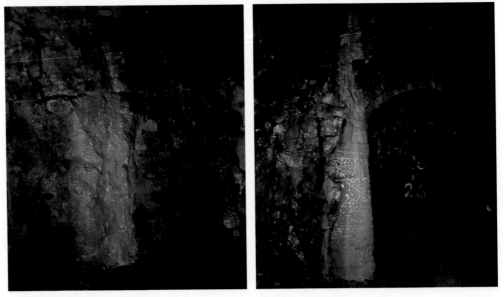

Figure 5.53 Wet drain-holes.

Figure 5.54 A brown long-eared bat *Plecotus auritus* roosting in plain sight in a railway tunnel drain-hole.

Figure 5.55 A serotine *Eptesicus serotinus* roosting in a recess that is off to the right of the back wall of the drain-hole. Looking in this wall pocket with a torch it just looked empty; unlocking it requires an endoscope with the lens bent at a right angle and turned inside the pocket like a key.

Photo 5.14 Breakdown below a 'textbook' stable domed ceiling.

Photo 5.15 An active choke in the entrance of a cavelet know to be exploited by greater horseshoe bats. Even the boulders and cobbles in the entrance are unstable, and a washing-machine-sized boulder in the middle of the entrance can be rocked with one finger of one hand: deadly dangerous and only assessable using remote-observation.

Figure 5.56 Waste stone and mixed sludge dumped in a worked-out cave (left) and an adit (right). The darkened patches on the lighter material are bat guano.

The enveloping PRF record anticipates the different features, allowing the relevant one to simply be circled. Thereafter the record encompasses: **1)** the PRF height above the floor in metres; **2)** the substrate the PRF is in; **3)** the recess entrance aspect (e.g. in the ceiling opening downward, or in a wall opening outward, etc.); **4)** the entrance dimension on the longest axis in centimetres; **5)** the entrance dimension on the shortest axis in centimetres; **6)** the bat species; **7)** the minimum number of bats visible; **8)** whether the bat is obviously awake or apparently torpid; **9)** whether there are any droppings in the PRF; and, **10)** where the bat is in relation to the entrance (i.e. above the entrance so the bat went in and up, in front of the entrance so it is easily viewed with a torch, to one side of the entrance so it went in an round, or below the entrance so it went in and down).

5.8 What 2.4: recording the advanced values for enveloping roost positions

The advanced values for enveloping roost positions comprise: **1)** the distance the bat is from the entrance; **2)** the internal height the roost feature extends above the entrance; **3)** the internal width the roost feature extends back from the entrance; **4)** the internal depth the roost feature extends below the entrance; **5)** whether a comprehensive inspection of the internal areas is possible; **6)** the visible humidity condition inside the roost feature; and, **7)** whether there are any associated invertebrates inside the roost feature.

5.8.1 The distance the bat is from the entrance

This value is recorded in centimetres and is most easily done using the endoscope. Obviously, the endoscope should not ever come into contact with the bat and ideally the intrusion will be as least disturbing as possible, so an estimate to the nearest 5 cm is a fair compromise.

5.8.2 The internal dimensions

The internal height is the apex of the roof of the PRF, i.e. the maximum that a bat could climb up inside the roost, and is taken from the apex of the entrance to the internal apex of the fissure or void.

The internal width is recorded from the middle of the entrance straight back on the horizontal to the back wall of the PRF. It is accepted that the width may be wider on the other axis (i.e. side-to-side rather than front-to-back) but the measurement is considered in respect of the bat accessing the PRF rather than occupying it once inside. In this context, the front-to-back dimension is more informative.

The internal depth is the floor of the PRF and taken from the base of the entrance.

5.8.3 Whether a comprehensive inspection was possible

This presents something of a paradox. For the answer to be yes, the 1st Recorder has to be confident that every bit of the internal area has been searched. In trees this is easy, but in rock roost features it is much more common for the 1st Recorder to finish with a nagging uncertainty.

After getting endoscope extensions jammed sufficiently often to have caused significant delays and expense, we now have a rule that if the inspection cannot be completed with a standard 1 m snake, we simply state that a comprehensive inspection was not possible. Experience has shown that attempting to negotiate corners inside a narrow crevice with 2 m+ of extension so rarely achieves a satisfactorily controlled result that it would be irresponsible to continue: the risk of damaging the PRF or injuring something occupying it is just too great.

> **Note:** Be careful when attempting to use a 17 mm lens because the head is so much wider than the snake and can get stuck if it passes through a hole in a narrow gap – when you come to withdraw that hole is suddenly so much harder to find because it is now invisible behind the lens, and the lens wants to go anywhere other than the hole. It is better to use a snake with a lens-head that is as close to the diameter of the snake as possible. You will still get it stuck but it causes less damage when you finally lose your temper and get both feet on the wall so you can rip the head off the thing …

5.8.4 Internal humidity

Tree-roost data have demonstrated that individual bat species display tolerances to different broad humidity regimes at different times of year. The basic classification is as follows:

1. **Dry** – arid with no suggestion of moisture;
2. **Damp** – the surface darkened by moisture but with no droplets or flow of water;
3. **Wet** – obvious beads, droplets, pooled and or flow of water.

5.8.5 Associated species

The associated species have already been described at Section 5.6.4.

5.9 The subterranean climate

> **Note:** This section refers to several accounts that were not listed in the introduction to the subterranean bat species given earlier in the chapter. This is because the environmental niche data include measurements taken in landforms such as cellars, which are not within the scope of this book, and also controlled investigations in laboratory studies. The studies listed in the introduction are limited to accounts of the specific subterranean landforms that are the focus of this book, but the environmental niche accounts are specific to the bat species regardless of where they were roosting, and these data are therefore relevant in this context.

> **Note:** This book specifically did not include built structures because houses, cellars, castles and even motorway bridges and high-rise blocks warrant their own book, and that book would be far better written by a different team of authors.

5.9.1 Introducing the invisible niche

It is easy to overlook the fact that even in the absence of the direct effect of the sun, the subterranean world has its own predictable climatic cycle. The climate offered is influenced by the seasons and prevailing weather on the surface, but also by the physical structure and subterranean extent of the individual landform, as well as water infiltration (Perry 2013).

The four broad subterranean landforms are different from each other, and individual examples of the same landform will have unique characteristics. Notwithstanding, some broad climatic commonalities can be accepted, as follows:

1. Contrary to what is still a popularly held belief, even in the absence of flowing water, the influence of daily outside weather conditions is felt below ground because it affects air currents which may extend hundreds of metres into subterranean systems (Cropley 1965).

2. Flowing water has a greater cooling effect than warming. In the spring, the rate of water flow will decrease and continue to diminish down to the minimum summer flow. As a result, there will be less warm water entering the system in the pregnancy, nursery and mating season than there was cold water entering through the autumn-flux, winter and spring-flux. The rate of warming is therefore slower and the magnitude of warming significantly less than the magnitude of cooling (Cropley 1965). In addition, if there is cold water entering at depth, there will be no warm air to rise and this will affect the air currents throughout the system cooling all the passages with direct connectivity to the surface, and as the rate of warming is slower and the warming effect less, it will take a longer time for the effect to pass upwards. By late mating season, the temperature should be approximating the mean annual surface temperature but it will quickly drop again in the autumn-flux with the increased flow of cold water (Cropley 1965).

3. A subterranean landform cannot be colder than the coldest temperature in the preceding 12 months, or hotter than the hottest temperature in the preceding 12 months. Although the relationship of annual mean surface temperature will be different in each individual system, a relationship will nevertheless **always** exist, which explains why caves in colder climates are cooler than those in warmer

climates; a cave in Yorkshire will **always** be cooler than a cave in Somerset. To really hammer the point home: if we dug up the cave in Somerset and translocated it to Yorkshire, it would be colder in its new location.

4. The warmest part of the landform in winter will typically be the coolest part in summer (e.g. de Freitas & Littlejohn 1987).

The earliest attempt to compartmentalise the subterranean environment that the literature review found is that of Cropley (1965). This splits the subterranean climate beyond the portal into three parts, as follows:

1. **The daily 'Zone 1' section** – referred to by Cropley as 'Zone 1'; this is the part of the landform that opens in from the surface climate and in which the subterranean climate is in tune with the daily surface climate. How far this Zone extends from the entrance will depend upon the characteristics of the landform (i.e. the amount of shade over the entrance(s); the size of the entrance(s); the number of entrances; the overall subterranean extent; the presence of flowing water, etc.).

2. **The semi-annual 'Zone 2' section** – referred to by Cropley as 'Zone 2'; this refers to all the parts of the landform that are sufficiently removed from the entrance that the daily climatic fluctuations experienced in Zone 1 are imperceptible, but in which seasonal variations in temperature and humidity still occur in response to air currents, the circulation of which changes from one half of the year to the next, i.e. in across the ceiling and out across the floor in summer, and in across the floor and out across the ceiling in winter.

3. **The annual 'Zone 3' section** – referred to by Cropley as 'Zone 3'; this refers to any parts that are so remote or otherwise removed by bends in the passages, or by being above the entrance, that no temperature variations occur to move the climate significantly away from the mean annual surface temperature. The annual section may experience air movement, and there may be changes in the direction of that air movement, but the velocity of that movement will be so low that any change in the direction of circulation will be functionally imperceptible.

Depending upon its size, an individual landform may have only Zone 1, or one or more Zone 1s connecting with one Zone 2 but no Zone 3s, or one or more Zone 1s which connect with one Zone 2 which opens into one or more Zone 3s, the latter of which may be interconnected or discrete from each other.

Poulson & White (1969) collated data from various sources to put together a table of typical environmental parameters offered at the portal and within the three subterranean zones inside caves in the temperate climate. Their tabulation included temperature and humidity ranges which are presented in a greatly simplified format in Figures 5.57 and 5.58. The tables illustrate how the range between the lowest and highest temperatures narrows through the zones, but it is recommended that readers read both Cropley (1965) and Poulson & White (1969) to get the full picture.

Research that began in the 1950s (e.g. Hooper & Hooper 1956; Bezem *et al.* 1964; Ransome 1968; Daan & Wichers 1968), and continued in the 1970s (e.g. Gaisler 1970; Ransome 1971; Daan 1973) demonstrates unequivocally that different bat species occupy different climates inside subterranean habitats. These original discoveries have been increased in detail since (e.g. Bogdanowicz & Urbańczyk 1983; Wermundsen & Siivonen 2010; Wermundsen 2010; Klys 2013) and all reach the same conclusion: each individual bat species occupies a subtly different niche on the climatic gradients in different months. This is what allows multiple species to coexist in the same space at the same time without conflict.

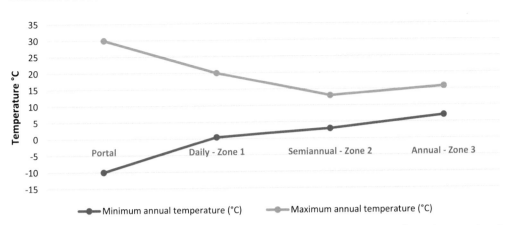

Figure 5.57 A general illustration of the reduction in the temperature ranges from the portal and across the three subterranean zones of a solution cave described by Cropley (1965) using data derived from Poulson & White (1969).

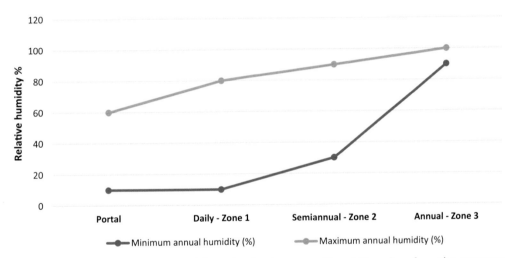

Figure 5.58 A general illustration of the reduction in range of humidity values from the entrance and across the three subterranean zones of a solution cave described by Cropley (1965) using data derived from Poulson & White (1969).

Daan & Wichers (1968) defined three climatic gradients they considered were the strongest influence on which bat species would occupy the individual situations in caves, Mihál & Kaňuch (2006) came to the same conclusion, and Klys (2013) later ranked them in order of significance as follows: **1)** the most important influence was air movement; **2)** the air movement influenced the temperature range and fluctuation within that range; and **3)** the air movement and temperature influenced the final factor: humidity.

However, the three are interlinked and a fourth factor is likely to determine bat distribution: **4)** illuminance, which will be a visual indication of how much solar radiation enters the system, and therefore how warm the air will be, and also how much light a bat perceives in different proportions of the system.

Each gradient has a range of values and each bat species occupies a niche position in association with a specific value on each individual gradient. Just like a combination lock, the right set of values opens the environment to the species, **BUT** this combination changes over the course of the annual cycle and how awake or asleep the species wants to be.

To illustrate, in an account of Finnish and Estonian systems, Natterer's bat occupied the coldest and driest conditions overall, and brown long-eared bats occupied colder and drier situations than Daubenton's bat and Brandt's/whiskered bats which occupied relatively warm and more humid situations (Wermundsen 2010). However, this situation varies as the hibernation season progresses. For example, in the autumn (October/November) Brandt's/whiskered bats, Daubenton's bats and brown long-eared bats all occupied warmer and more humid situations than they did in spring (March/April).

It is vital to bear in mind that bats are not primary hibernators that enter torpor at the imposition of the climate; they are more advanced than that and their mobility and metabolism offer them the facility to use torpor at will by choosing to roost in a specific situation.

Ransome (1971) collated and reviewed historical and contemporary literature and considered it in light of his own detailed observations of greater horseshoe bats which demonstrated that the longest individual bats remained in torpor without changing position was 10 days. The perception is that bats of all species use torpor as a tool to be applied to span time, in the same way that we might use breathing apparatus to span a sump. Individual species are able to select their optimum hibernation environmental niche by altering the distance they roost from the entrance of the subterranean system, and also the height of their roost position from the floor (Kokurewicz 2004).

> **Note:** Investigations of the cave environment take the broad assumption that caves are only occupied in the cooler half of the year: October through April. As a result, there is a possibility that the reported data and accounts are incomplete. This presents an opportunity for useful and interesting studies, particularly in the case of the enveloping PRF.

5.9.2 What we need to know to use the knowledge to inform surveys of new sites and monitoring of artificial landforms

The biggest problem we have when designing a survey of a subterranean landform is knowing where to direct survey effort in both physical/visual terms (i.e. where to look) and in terms of temporal scope (i.e. when and how often).

Over repeated visits the surveyor becomes more familiar with a system and they tend to find that the bats that roost in exposed situation PRF are found in some areas and not others. The same is true of enveloping PRF: they are condensed into specific parts of the system between wide areas of barren rock.

The larger the landform, the longer it takes to find and map the different situations. This is further frustrated by the fact that the bats are not all in the system in all months of the year. In addition, even when an individual species is present in the system over multiple months, it is not sedentary within the system, but moves around in response to invisible cues. As we have seen, this is not just our perception, it is a phenomenon which Daan & Wichers (1968) termed 'internal migration'.

Daan & Wichers (1968) showed that, while the barbastelle, Natterer's bat and long-eared bats remained within the Portal and Daily (Zone 1) section throughout the period they occupied the cave system, the small myotis and Daubenton's bats favoured the Annual (Zone 3) section in the autumn-flux period and gradually moved forward into the Semi-annual (Zone 2) section as winter progressed into the spring-flux period.

Finally, the lesser horseshoe bat favoured the Semi-annual (Zone 2) section and Annual (Zone 3) sections throughout the period they occupied the cave.

A study performed by Mihál & Kaňuch (2006) is notable in that they found that air circulation, ambient temperature and relative humidity were the primary influences upon the occupancy of a subterranean system, and that: **1)** the altitude; **2)** the orientation of the entrance; **3)** the size of the entrance; and, **4)** the presence of standing water, had no significant influence on which bat species would be present, or the numbers of bats present. While they found that the overall length of the system did have a significant positive effect, this was only inasmuch as it meant that more combinations of airflow, temperatures and humidity were offered by bigger systems. However, even where a system was shorter and typically colder, that did not mean it was universally bad: both barbastelle and serotine were positively associated with these situations.

A landform offering all three Zones beyond the Portal may be sufficiently large for it to take a team of surveyors several days to adequately catalogue all the different roost positions. And, as the roost positions may change from one month to the next, it may be that even mapping all the easily accessible roosts positions takes an entire year, and even several years in succession.

But even when we have recorded all the occupied roost positions, how do we know that the positions the bats were not recorded in are unsuitable? How do we know that the places that were not occupied, might not be occupied in a successive year? How do we know our map is definitive?

This problem extends beyond the survey into our attempts to deliver compensation or enhancements. How do we assess whether what we have made will work? Suppose we create an artificial cave: how do we demonstrate success? Suppose no bats adopt it in the first year: how do we assess why that is? Suppose the wrong bat species adopts it: how do we assess why the target species has not? And how long do we keep up surveillance: one year, ten years, 100 years?

If we know that the climate is correct, we can direct resources to ensure there is no barrier to the target species finding and adopting the landform. Once we have achieved the correct climate, and ensured connectivity, we can widen our surveillance interval to give the bats time to find it before we run out of money and goodwill.

So how do we know the environment is correct and that the reason the target species has not colonised it must be as a result of an impact outside? How do we stop our surveys from being 'Ecochaeology',[44] and our attempts at compensation and enhancement from being roulette? How do we learn and progress?

The answer is that we identify the niche occupied on the airflow, temperature and humidity gradients by our target species, and we design surveillance to find that niche, and monitoring to deliver that niche.

In the subsections that follow, the detailed investigations performed by a wide range of top-class scholars are collated and presented to provide: **a)** the period in which each bat species exploits subterranean landforms; **b)** the overall tolerance range occupied by each bat species on the three gradients; **c)** the general niche position occupied within that range; and, **d)** (for some species) the niche position occupied in individual months of the year.

Armed with this information, we can deploy temperature and humidity loggers and also manually assess the climate and compare the data we gather against the thresholds occupied by our target species.

We begin by assessing whether the environment offered by our landform is within the overall range tolerated by our species. If it is not, we will know that alteration is required, and we will also know what the alteration needs to deliver in terms of airflow, temperature, humidity and illuminance; we therefore have a sense of purpose and can move forward to secure that first small victory.

When we are confident the climate remains within the tolerance range throughout the period the species typically exploits the landform, we can proceed with assessing whether it delivers the favoured niche. The tolerance spans the entirety of the occupancy period, but the niche position within that range may change, so we need detailed data across individual months.

With this information, we can make incremental alterations to the physical environment to create the correct climate: the invisible niche. This may take years to deliver, each alteration requiring another year of monitoring to assess its effect. However, once delivered the niche is permanent and all the effort will be worthwhile.

At the very least, delivering the climatic niche gives the project a measurable objective.

5.9.3 Airflow

Over a century ago Crammer (1899) found that a cave with one entrance at the bottom might trap warm air in the summer, and a cave with one entrance at the top might trap cold air in the winter. In the latter group, the effect was that in winter the warmer air in the cave rose, passing out across the cave ceiling, and drawing in more cold air down over the cave floor. The following summer the air outside was warm, so the cold air trapped in the cave remained where it was. Although there was some warming, Crammer (1899) deduced that cold air traps are the coldest caves year-round. The effect is reversed for warm air traps, which are consequently the warmest caves year-round. Regardless, in these caves airflow is imperceptible.

If there were two entrances at different altitudes, however, there is airflow because of the difference in air density between the air inside and outside the cave (Crammer 1899). The flow passes from the higher to the lower entrance in summer and the opposite in winter (Crammer 1899). Because humid air weighs less than dry air the airflow is increased in caves with water-filled sections (Crammer 1899).

Cullingford (1962) noted that airflow velocity was also associated with free surface streams in active systems, and that air was carried along in the direction of the water. An example is given of a stepped pothole which a stream flows into and ultimately disappears down a flooded sump. Air is drawn down with the stream, but returns from the flooded sump, passing over the ceiling and exiting as a return draft via the same swallow-hole through which the stream enters. This is illustrated in Figure 5.59.

In at least one case, return air currents at a potential surface entrance have led to the discovery of the system below (Cullingford 1962).

Daan & Wichers (1968) also found that in the warmer half of the year (i.e. the pregnancy, nursery and mating periods for bats) caves have a chilling influence on air that passes through a cave system which results in generally downward travel of warm air, but this is reversed in the cooler part of the year (i.e. the autumn-flux, winter and spring-flux periods) with warm air moved in generally upward travel.

Daan (1973) highlighted that the presence of a shaft increased airflow. This is also the case if there is more than one entrance at different heights (e.g. Penn Recca mine (see Edwards 2011)). The ambient temperature threshold, above or below which the direction of airflow was determined, was reported to be 4–6 °C (Daan 1973). When the temperature is above this the airflow enters across the ceiling and exits across the floor, with small *Myotis* spp. and Daubenton's bats favouring the areas furthest from the threshold. Below this threshold the airflow enters across the floor and exits across the ceiling and small *Myotis* spp. and Daubenton's bats move to positions in the Daily (Zone 1) section.

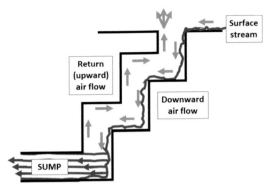

Figure 5.59 Closed circulation of air in response to stream flow (taken from Cullingford 1962).

Wind may also contribute to the flow velocity if there is more than one entrance to the system and they are at different altitudes, aspects and orientations and therefore not subject to the same gusts of wind at balanced equal power. The direction of wind also has an effect on the air movement patterns, i.e. where the wind blows into one entrance of a system it will exit via any other entrance that is not subject to the same gusts, but where the wind blows across an entrance rather than blowing into it, there is a drawing effect, with air inside the system sucked towards the entrance across which the wind is blowing (Cullingford 1962).

The final factor is barometric pressure: Cullingford (1962) noted that even in the absence of wind a change in air pressure influenced subterranean air movement, with low pressure resulting in an outward draught and high pressure resulting in an inward flow.

However, de Freitas et al. (1982) had the last word when they proved that the direction of airflow was as a result of a substantial difference between the mean density of air inside and outside the subterranean system. Providing this remained above and below the critical threshold the direction would be constant across months. The difference between the summer downflow and the winter up-flow is therefore a real and transferable phenomenon, and continues night and day through those periods. However, in the transition periods of the spring-flux and autumn-flux, there are occasional weak daily reversals, with a daily up-flow but nightly downflow in the spring-flux, and a daily downflow but nightly up-flow in the autumn-flux. Notwithstanding, the airflow routes are similar in both summer and winter (de Freitas et al. 1982).

Vandel (1965) was the first to consider the effect of air movement upon different faunal groups in any detail. He found that a system that was subjected to a permanent air current with no places in which faunal organisms might hide, was usually azoic (i.e. barren). However, any situation in that cave in which an organism might shelter out of the draught would be exploited.

With regard to bat species, two studies in the Netherlands have reported that although bats of all species avoid being in direct draught flow, Vespertilionid species do nevertheless occupy enveloping PRF that open onto draughty corridors (van Nieuwenhoven 1956, Daan & Wichers 1968).

The pattern and range of airflow through the Daily (Zone 1) section and Semi-annual (Zone 2) section influence the distance individual bat species will roost in a specific landform, but another factor may be at play – the wetting/drying effect of those flows.

de Freitas & Littlejohn (1987) found that in the winter the air not only cooled the cave ceiling as it passed across it, it also significantly dried it. As a result, where in the summer the airflow resulted in condensation on the ceiling and upper walls, the winter was characterised by a strong evaporation effect and the drying of the ceiling and walls. This may explain why bats of all species are found in exposed positions on the ceiling and walls in the mating and early part of the autumn-flux periods, but appear to seek recessed and enveloped situations, and often lower down, through the latter part of the autumn-flux and winter periods.

Thus, all open subterranean rock landforms are subject to the same patterns of airflow, but different landforms will exhibit different velocities. de Freitas *et al.* (1982) list the influences of airflow velocity as including: **1)** the number of entrances; **2)** elevation distances between entrances; **3)** passage geometry and configuration; **4)** presence of flowing water and entry situation (i.e. surface entry or subterranean emergence); resonance of air in chambers and caverns; **5)** funnelling of wind; and, **6)** changes to water levels in flooded sections (and therefore the extent of air-filled passage).

Klys (2013) measured airflow using an anemometer in a mine system that encompasses over 300 km of subterranean habitat. Each measurement was taken for 1 minute at a position 1.2 m above the floor. The conclusion was that in enveloped roost positions, Brandt's bat, whiskered bat, Natterer's bat and brown long-eared bat could tolerate an airflow of 1 m/s, but Daubenton's bat was least evident in situations where airflow was recorded. This mirrors the results of Kokurewicz (2004) who recorded Daubenton's bats occupying situations where the airflow was less than 0.04 m/s.

Bogdanowicz (1983) and Bogdanowicz & Urbańczyk (1983) found Daubenton's and Natterer's bats favoured situations where they were isolated from the effect of the external atmospheric conditions. In contrast, barbastelles and brown long-eared bats were found in situations where airflow was greatest and individuals were recorded in 'draughty' conditions in which 'there were considerable variations in hygrothermal conditions'. The same situation was reported by Ransome (1971), who found that greater horseshoe bats favoured a degree of air movement: when air vents were closed in a cellar the number of bats fell and when the vents were reopened the numbers increased. Upon investigating further, he found the airflow influenced temperature fluctuations and therefore delivered the correct niche value on the temperature gradient.

The take-home message is that different species of bats have different tolerances to airflow, and some of them may exploit the airflow as a signal that conditions outside the system are suitable for them to be able to effectively forage; so, they can go into torpor secure in the knowledge that the airflow will wake them if there is a useful climate window.

The issue is that exactly what degree of airflow each species can tolerate and favours within that range is unknown: if we want the data we are going to have to go and get it. So here is our first question: what metres-per-second (m/s) airflow will each bat species tolerate?

However, if we are to answer this question, it is vital to take the measurement **BOTH** at the bat and in the middle of the passage because the airflow in the two situations may be very different. Figure 5.60 illustrates the sheltering effect of a rough surface using a beam of light, demonstrating that a rock face typically has indentations in it. These may not seem that deep to a person, but to they are very deep relative to the body of a bat, and the equivalent of us lying in a dyke that is several metres deep. These wall and ceiling pockets will certainly offer shelter from the wind to a bat.

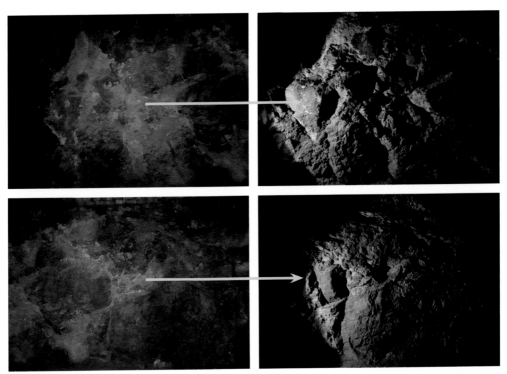

Figure 5.60 Left: view of a railway tunnel wall with light shining perpendicular onto it. Right: the same view of a railway tunnel wall with light shining parallel down it.

Note: Anyone wishing to look at airflow in greater detail might like to read the following review:

- Perry, R. 2013. A review of factors affecting cave climates for hibernating bats in temperate North America. *Environ. Rev.* 21: 28–39.

Although the title says North America, the information is applicable to caves everywhere.

However, another three papers simply do not get the credit that they deserve and should always be cited in descriptions of the airflow dynamic:

- Crammer, H. 1899. Eishöhlen-und Windröhren Studien. *Vienna Geographischen Gesellshaft* 1: 19–76 – Crammer was the first to describe the phenomenon but is overlooked by many papers
- Cropley, J. 1965. Influence of surface conditions on temperatures in large cave systems. *Bull. Nat. Speleol. Soc.* 27: 1–10 – Cropley was the first to identify and describe the three zones; again, often overlooked.
- Poulson, T. & White, W. 1969. The cave environment: limestone caves provide unique natural laboratories for studying biological and geological process. *Science* 165(3897): 971–981 – interesting collation of data in an attempt to predict the subterranean climate across temperate caves (and we do so love our predictive intelligence frameworks …)

After that the various collaborations of the late Chris de Freitas working in Glowworm Cave are a masterwork in cave climatology, but also lighting the way in what consultants really should be considering in environmental impact assessments.

5.9.4 Temperature

While accepting that the vegetation over the cave has an influence, Vandel (1965) notes that caves become colder with increasing altitude. The effect is described as decreasing by 2–3 °C for each 100 m increase (Vandel 1965). In the opposite direction, Vandel (1965) notes that it is rare to find a regular and predictable linear increase in temperature in relation to depth in solution caves, but Hayman (2016) suggests that the temperature in a coal mine was reckoned to rise by 1 °C for every 40 m depth a colliery shaft descended. In summary, the higher the system is above sea level the colder it is, but there is not a predictable increase in temperature brought about solely by virtue of the depth of a location below ground level.

Webb *et al.* (1996) produced a comprehensive review of temperature niches occupied by all the bat species native to the UK and many more besides (if they had done the same thorough job with airflow and humidity this chapter would be half the size).

A theme common to every scientific study is that different species of bats tolerate different temperature ranges and even where the ranges overlap, there are differences in the niche positions occupied. True, some of the niche positions have overlaps across pairs of species, but the fundamental message is that you cannot apply a 'one-size-fits-all' approach to subterranean roosting any more than you can to roosts in any other situation. Even in winter, the differences are more evident than the commonalities.

Mihál & Kaňuch (2006) found the *Vespertilionidae* had significantly greater tolerance to lower temperatures than the *Rhinolophidae*. Within that broad division, Bogdanowicz & Urbańczyk (1983) found Daubenton's and Natterer's bats in stable roost climates, barbastelles and brown long-eared bats were more tolerant of variation in temperature (and humidity). However, only Ransome (1971) has considered daily temperature fluctuations (i.e. what greater horseshoe bats could tolerate in a single 24-hour period without waking).

Ransome (1971) found that although most bats were recorded in situations where the temperature was 7–9 °C, the average temperature selected by an individual bat might vary between 5 °C and 12 °C, and below the 5 °C limit the bats temporarily abandoned the roost. Notwithstanding, the daily tolerance for temperature fluctuations was high: between November and February the bats tolerated a daily fluctuation of up to 2.7 °C, and this increased from March through May to a maximum fluctuation across 3.4 °C in a single 24-hour period.

In summary, providing the temperature remains above 5 °C greater horseshoe bats can tolerate daily fluctuations in temperature across 3.4 °C without being disturbed.

Although the tolerance for daily fluctuations is not reported for other species, data are available that describe the broad temperature range some bat species can tolerate and also the niche temperature position occupied within the overall range. These data are presented in Tables 5.9–5.12.

> **Note:** The data presented in Tables 5.9–5.12 have a wide geographical spread and encompass countries with very different climates from the British Isles: e.g. British Isles (Ransome 1968), Holland (Daan & Wichers 1968), Czech Republic (Gaisler 1970), Poland (Kowalski 1953; Bogdanowicz & Urbańczyk 1983), Finland and Estonia (Wermundsen 2010; Wermundsen & Siivonen 2010).
>
> These data are provided to illustrate niche positioning and might be used to construct a hypothesis to test in the British Isles, but should be used with caution in any predictive assessment.

Table 5.9 Temperature range (°C) tolerated by Vespertilionid species (i.e. simple-nosed Vesper bats) in subterranean systems.

SPECIES	TEMPERATURE TOLERANCE RANGE °C					
	Kowalski 1953 (unknown)	Harmata 1969 (unknown)	Daan & Wichers 1968 (December – March)	Gaisler 1970 (January – February)	Bogdanowicz & Urbańczyk 1983 (October – April)	Lesiński 1986 (October – April)
Barbastelle *Barbastella barbastellus*	0–4	–3–9	—	0–6	–3–4.8	–2.5–6.5
Serotine *Eptesicus serotinus*	—	0.5–6	—	2–4	—	0.5–6.5
Bechstein's bat *Myotis bechsteinii*	—	—	—	1–7	—	—
Brandt's bat *Myotis brandtii*	—	—	—	2–8	—	—
Daubenton's bat *Myotis daubentonii*	—	—	2.2–10.4	3–8	0.2–10	0.5–7
Whiskered bat *Myotis mystacinus*	2–4	—	0–10.5	2–8	—	—
Natterer's bat *Myotis nattereri*	—	6–10	3.3–8.1	3–7	–2.1–8	1.5–7
Brown long-eared bat *Plecotus auritus*	0–7 *Plecotus* sp.	–3–11	0–5.3 *Plecotus* sp.	2–7	–0.8–9	1–1.5
Grey long-eared bat *Plecotus austriacus*		—		2–6	—	–0.5–4.5

Table 5.10 Temperature range (°C) tolerated by Rhinolophid species (i.e. horseshoe bats) in subterranean systems.

SPECIES	TEMPERATURE TOLERANCE RANGE °C					
	Kowalski 1953 (unknown)	Ransome 1968 (October – April)	Harmata 1969 (unknown)	Daan & Wichers 1968 (December – March)	Gaisler 1970 (January – February)	Harmata 1987 (October – April)
Lesser horseshoe bat *Rhinolophus hipposideros*	6–7	—	2–14	6.4–10.4	4–12	2–10
Greater horseshoe bat *Rhinolophus ferrumequinum*	—	3.5–12.5	—	—	7–12	—

Table 5.11 Niche positions occupied by Vespertilionid species (i.e. simple-nosed Vesper bats) on the temperature gradient.

SPECIES	TEMPERATURE NICHE °C					
	Daan & Wichers 1968 (October – April (no March value))		Harmata 1969 (unknown)	Bogdanowicz 1983; Bogdanowicz & Urbańczyk 1983 (October – April)	Wermundsen & Siivonen 2010; Wermundsen 2010 (October – April)	
Barbastelle *Barbastella barbastellus*	Dec	0	4	0–3	—	
Brandt's bat *Myotis brandtii*	—		—	—	Oct–Nov	6.2 (±2)
					Dec–Feb	4.5 (±2.2)
					Mar–Apr	4.2 (±2)
Daubenton's bat *Myotis daubentonii*	Oct	9.2	—	1.5–6	Oct–Nov	6.5 (±2.6)
	Nov	8.9				
	Dec	7			Dec–Feb	4 (±1.6)
	Jan	6.8				
	Feb	6.5				
	Apr	6.4			Mar–Apr	3.7 (±1.8)
Whiskered bat *Myotis mystacinus*	Oct	10.1	—	—	Oct–Nov	6.2 (±2)
	Nov	8.9				
	Dec	6.6			Dec–Feb	4.5 (±2.2)
	Jan	5.9				
	Feb	6.2				
	Apr	6.2			Mar–Apr	4.2 (±2)
Natterer's bat *Myotis nattereri*	Oct	—	—	2–6.5	—	
	Nov	—				
	Dec	6.4				
	Jan	6.8				
	Feb	5.9				
	Apr	5.6				
Brown long-eared bat *Plecotus auritus*	Oct	—	6	0.5–4	Oct–Nov	5.8 (±2.8)
	Nov	—				
	Dec	5.2			Dec–Feb	3.2 (±2)
	Jan	0				
	Feb	5.3				
	Apr	—			Mar–Apr	3.4 (±1.2)

Table 5.12 Niche positions occupied by Rhinolophid species (i.e. horseshoe bats) on the temperature gradient.

SPECIES	TEMPERATURE NICHE °C				
	Ransome 1968 (October – April)		Daan & Wichers 1968 (October – April (no March value))		Harmata 1969 (unknown)
Lesser horseshoe bat *Rhinolophus hipposideros*	—		Oct	10.4	7–8
			Nov	9.7	
			Dec	8.2	
			Jan	8.3	
			Feb	7.9	
			Apr	7.5	
Greater horseshoe bat *Rhinolophus ferrumequinum*	Oct	10.5 (±1.1)	—		—
	Nov	9.8 (±1.3)			
	Dec	8.4 (±1.4)			
	Jan	7.6 (±1.6)			
	Feb	6.6 (±1.6)			
	Mar	7.4 (±1.5)			
	Apr	7.5 (±1.2)			

The data are presented in Tables 5.9–5.12 for practical use. They have been collated in order that the reader may enter them into a spreadsheet or a report for comparison with their own data. However, they are very 'dry' in this format so the monthly mean values provided by Daan & Wichers (1968) and Ransome (1968) have been used to create a visual illustration of the temperature niches occupied by Daubenton's bats, whiskered bats, Natterer's bats, long-eared bats, lesser horseshoe bats and greater horseshoe bats, and how those niches change in each month with a general descending trend (see Figure 5.61). The decreasing niche temperature trend is also visible in the Wermundsen & Siivonen (2010) data, as illustrated in Figure 5.62.

Looking at the temperature data presented in Figures 5.61 and 5.62, the question is: did the temperature in the position the bats were occupying change and the bats remain in it, or did different roost positions offer different temperatures and the bats moved between them, or did the temperatures move around in the system and the bats moved with them.

To clarify, for example: the greater horseshoe bat data recorded by Ransome (1968) shows the bats occupying 10.5 °C in October and 9.8 °C in November, but was there always a position offering 10.5 °C and after October the bats did not occupy it, or was that temperature not available and 9.8 °C was the next best thing?

5.9.5 Humidity

In fossil caves and dry mines humidity is the result of warm air being drawn from the outside and into the subterranean situation where it is cooled on the surfaces hidden from the sun. In summer, the humid air of warm nights is cooled as it is drawn into the cave, so the air within becomes saturated (Chapman 1993). As the air temperature falls below the dew-point, condensation occurs and the air becomes increasingly humid. The effect is most visible on the roof of the system, which may become sufficiently damp for droplets to reflect light from below and to drip continuously (Cullingford 1962).

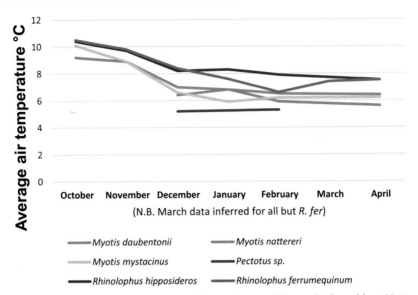

Figure 5.61 Average temperatures occupied by Daubenton's bat, whiskered bat, Natterer's bat, long-eared bat and lesser horseshoe bat from October through April (data taken from Daan & Wichers 1968; N.B. March value inferred for illustrative purpose), and greater horseshoe bat (data taken from Ransome 1968).

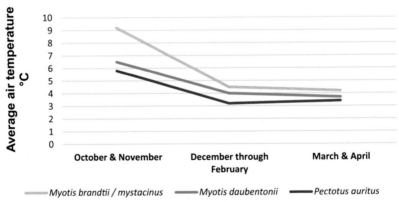

Figure 5.62 Average temperatures occupied by small Myotis bats, Daubenton's bats and brown long-eared bats in lumped periods: October and November, December through February, and March and April (data taken from Wermundsen & Siivonen 2010).

The reverse is true in winter, when the air being sucked in is cooler and is warmed by contact with the walls. This incurs a saturation deficit which, if large enough, may have a desiccation effect on the roost substrate as a result of evaporation, and this in turn results in increased cooling (Cullingford 1962, Chapman 1993). If the system has a large volume and/or large vertical distance between entrances the airflow is increased and the drying is significantly higher and the humidity significantly lower than it is in summer (Chapman 1993). This may be less of an issue if the cave has numerous indentations in the ceiling, because water vapour is lighter than air, and therefore rises to become trapped (Chapman 1993).

The point is that a cold cave will not necessarily be a damp cave, but that does not automatically mean it is a bad cave from the perspective of a bat: remember – different bat species are looking for different environments and there is no place for sweeping generalisations in ecological assessments. As with everything in nature, requirements for humid sites are on a sliding scale with the greater horseshoe bat at one end, favouring a relative humidity in excess of 90% (Ransome 1971; Jones *et al.* 1995), Daubenton's bats somewhere in the middle and doing best in environments with *c.*80% relative humidity (Daan 1973), and the serotine at the other end, requiring a relatively low humidity[45] (Schober & Grimmberger 1993). As with the temperature gradient, only Ransome has described fluctuations in humidity across the tolerance range with one greater horseshoe bat torpid for ten days, during which it tolerated a relative humidity which varied from 84% to 95%.

Although (as with the temperature data) the tolerance for daily fluctuations in humidity is not reported for other species, data are available that describe the broad range some bat species can tolerate and also the niche humidity position occupied within the overall range. The data are presented in Tables 5.13–5.16.

> **Note:** As with the temperature data, the humidity data presented in Tables 5.13–5.16 have a wide geographical spread and encompass countries with very different climates from the British Isles: e.g. Poland (Bogdanowicz & Urbańczyk 1983), Finland and Estonia (Wermundsen 2010; Wermundsen & Siivonen 2010).
>
> These data are provided to illustrate niche positioning and might be used to construct a hypothesis to test in the British Isles, but should be used with caution in any predictive assessment.

As with temperature, the part of the cave that offers the correct humidity may be different in different months and even in response to short-term weather conditions.

Punt & Parma (1964) originally described a 'condensation-zone' that might be thought of as a 'Goldilocks zone' for cave-roosting bats. This idea was enlarged by Daan & Wichers (1968) who suggested the phenomenon as a potential explanation for why the

Table 5.13 Humidity range (% rh) tolerated by Vespertilionid species (i.e. simple-nosed Vesper bats) in subterranean systems.

SPECIES	HUMIDITY TOLERANCE RANGE % rh	
	Bogdanowicz & Urbańczyk 1983 (October – April)	Lesiński 1986 (October – April)
Barbastelle *Barbastella barbastellus*	60–99	72–98
Serotine *Eptesicus serotinus*	—	78–92
Daubenton's bat *Myotis daubentonii*	71–100	80–98
Natterer's bat *Myotis nattereri*	73–100	85–100
Brown long-eared bat *Plecotus auritus*	54–100	75–95
Grey long-eared bat *Plecotus austriacus*	—	69–90

Table 5.14 Humidity range (% rh) tolerated by Rhinolophid species (i.e. horseshoe bats) in subterranean systems.

SPECIES	HUMIDITY TOLERANCE RANGE % rh
	Ransome 1971 (November – May)
Greater horseshoe bat *Rhinolophus ferrumequinum*	76–100

Table 5.15 Niche positions occupied by Vespertilionid species (i.e. simple-nosed Vesper bats) on the humidity gradient. N.B. The Finish and Estonian data are presented as averages across two/ three months and also as averages across the entire 'winter period' in exposed and enveloped roost positions.

SPECIES	HUMIDITY NICHE % rh				
	Bogdanowicz 1983; Bogdanowicz & Urbańczyk 1983 (October – April)	Wermundsen & Siivonen 2010 (October – April)		Wermundsen 2010 (October – April)	
Barbastelle *Barbastella barbastellus*	83 (±7)	—		—	
Brandt's bat *Myotis brandtii*	—	Oct–Nov	91 (±14)	Exposed	89 (±15) 97 (±9) 94 (±8)
		Dec–Feb	89 (±13)	Enveloped	81 (±15) 92 (±5) 93 (±9)
		Mar–Apr	86 (±9)		
Daubenton's bat *Myotis daubentonii*	91 (±6)	Oct–Nov	93 (±16)	Exposed	92 (±23) 94 (±8) 96 (±10)
		Dec–Feb	91 (±9)	Enveloped	86 (±21) 91 (±14) 94 (±16)
		Mar–Apr	87 (±12)		
Whiskered bat *Myotis mystacinus*	—	Oct–Nov	91 (±14)	Exposed	89 (±15) 97 (±9) 94 (±8)
		Dec–Feb	89 (±13)	Enveloped	81 (±15) 92 (±5) 93 (±9)
		Mar–Apr	86 (±9)		
Natterer's bat *Myotis nattereri*	94 (±5)	—		Exposed	—
				Enveloped	88 (±6) 95 (±6)
Brown long-eared bat *Plecotus auritus*	85 (±10)	Oct–Nov	92 (±19)	Exposed	86 (±20) 92 (±14) 94 (±12)
		Dec–Feb	91 (±11)	Enveloped	85 (±20) 89 (±15) 91 (±11)
		Mar–Apr	84 (±9)		

Table 5.16 Niches positions occupied by Rhinolophid species (i.e. horseshoe bats) on the humidity gradient.

SPECIES	HUMIDITY NICHE % rh
	Jones *et al.* 1995 (month(s) unknown)
Greater horseshoe bat *Rhinolophus ferrumequinum*	>96

greater proportion of bats in their study were recorded in the deep periphery of the cave system in the autumn-flux period, but this altered as the temperature dropped through the winter. Their perception was that in the autumn-flux the air movement dried the ceiling and walls while dampening the cave floor. As a result, the majority of the bats remained in the deepest parts of the cave where the walls and ceiling were most humid. As the temperature dropped, so the air circulation reversed, dampening the walls and ceiling, which appeared to make these situations more attractive to the bats. The effect was to move the Goldilocks zone forward and ultimately into the cave threshold.

Note: Whether standing or flowing water has a positive or negative influence on the presence of specific bat species is still unclear, but the perception in some accounts is that flowing water has a negative influence and standing water does not. To illustrate:

Glover & Altringham (2008), found that swarming was significantly less likely to take place in the entrance of caves that held flowing water and, although Edwards (2011) did find roosting bats in Penn Recca Slate Mine (species unknown), he perceived an aversion to flowing water: 'In the Combe adit, bats were often seen hanging from the western wall, but never over the stream that flows on the eastern side.' However, Nagy & Postawa (2010) reported that in Romania the barbastelle, noctule and common pipistrelle hibernated in large and cold cave systems with a constant flow of water.

In the case of standing water, Bogdanowicz (1983) reported '*M. daubentonii* [Daubenton's bats] preferred sheltering places flooded with rainfall'. Mihál & Kaňuch (2006) found that the presence of standing water had no significant influence on which bat species would be present, or the numbers of bats present. And, in 2021 records were made of three greater horseshoe bats roosting in a mine adit and directly above water that was 50 cm deep (*BRHK Rock Face Database*).

Another aspect that is worthy of consideration is flooding, which would render areas of a natural cave or mine hazardous for roosting, but also alter the environment in the areas above the flood level as the extent of the system was decreased in line with the rise in water and then increased again as the water subsided. No reference to the effect of this phenomenon on the subterranean environment was found in any publication.

5.9.6 Illuminance

And so, at last we come to illuminance, the amount of light a bat perceives in its roost position. This is an easy section to write because the literature review found no structured data at all and just as we got the kit to go and get a data set together, COVID-19 prevented the project from going into the immediate proximity of the bats.

5.10 The subterranean rock bat species summaries

The data available in 2021 are insufficient to create a comprehensive account for any species (if they were not, there would not be any need for this project). However, a cursory account is provided to give an environmental baseline.

All the accounts are constructed from the sources cited earlier in the chapter. The roost exposure figures are constructed by collating data presented by Bezem *et al.* (1964), Daan & Wichers (1968), and Bogdanowicz & Urbańczyk (1983). Temperature and humidity ranges are the upper and lower values cited in Section 5.9.

N.B. All summary data are taken to include all studies provided in tables in earlier sections of the report unless stated otherwise.

Table 5.17 Barbastelle *Barbastella barbastellus* subterranean rock roosting account.

MACRO											
Landforms exploited	Solution cave ✓		Sea cave ?				Mine ✓			Railway tunnel ✓	
Months in which subterranean landforms occupied	Winter		Spring-flux		Pregnant		Nursery		Mating		Autumn-flux	
	Jan ✓	Feb ✓	Mar ✓	Apr ?	May ?	Jun ?	Jul ?	Aug ?	Sep ?	Oct ?	Nov ?	Dec ✓

MESO	
Roost situation	Threshold ✓
	Dark-zone ?

Roost exposure	
	Values taken from Daan & Wichers (1968) and Bogdanowicz & Urbańczyk (1983)

MICRO	
Roost height (m)	Range: 0.4–6 m
	Niche: ?
Temperature (°C)	Range: −3–9 °C
	Niche:* Oct–Apr: 0–3 °C
Humidity (% rh)	Range: 60–99%
	Niche:* Oct–Apr: 83% (±7)
Associates	Brown long-eared bat *Plecotus auritus*

* Bogdanowicz 1983; Bogdanowicz & Urbańczyk 1983.

Table 5.18 Serotine *Eptesicus serotinus* subterranean rock roosting account.

MACRO											
Landforms exploited	Solution cave ✓		Sea cave ?		Mine ✓			Railway tunnel ✓			
Months in which subterranean landforms occupied	Winter		Spring-flux		Pregnant		Nursery		Mating		Autumn-flux	
	Jan ✓	Feb ✓	Mar ✓*	Apr ?	May ?	Jun ?	Jul ?	Aug ?	Sep ?	Oct ?	Nov ?	Dec ?

MESO	
Roost situation	Threshold ✓
	Dark-zone ✓

Roost exposure	Roost position exposure: Exposed, Recessed, Enveloped; % of roosts (0–100). Values taken from Bogdanowicz & Urbańczyk (1983)

MICRO	
Roost height (m)	Range: 2.5–5 m
	Niche: ?
Temperature (°C)	Range: 0.5–6.5 °C
	Niche: ?
Humidity (% rh)	Range: 78–92%
	Niche: ?
Associates	?

Table 5.19 Bechstein's bat *Myotis bechsteinii* subterranean rock roosting account.

MACRO											
Landforms exploited	Solution cave ✓		Sea cave ?			Mine ✓			Railway tunnel ✓		
Months in which subterranean landforms occupied	Winter		Spring-flux		Pregnant		Nursery		Mating		Autumn-flux	
	Jan ?	Feb ?	Mar ?	Apr ?	May ?	Jun ?	Jul ?	Aug ?	Sep ?	Oct ?	Nov ?	Dec ?
MESO											
Roost situation	Threshold ✓					Dark-zone ✓						
Roost exposure	?											
MICRO											
Roost height (m)	Range: ?											
	Niche: ?											
Temperature (°C)	Range: 1–7 °C											
	Niche: ?											
Humidity (% rh)	Range: ?											
	Niche: ?											
Associates	?											

Table 5.20 Brandt's bat *Myotis brandtii* subterranean rock roosting account.

MACRO											
Landforms exploited	Solution cave ✓		Sea cave ?		Mine ✓			Railway tunnel ✓			
Months in which subterranean landforms occupied	Winter		Spring-flux		Pregnant		Nursery		Mating		Autumn-flux
	Jan ✓	Feb ✓	Mar ✓	Apr ✓	May ✓	Jun ?	Jul ?	Aug ?	Sep ?	Oct ✓	Nov ✓	Dec ✓

MESO		
Roost situation	Threshold ?	Dark-zone ✓
Roost exposure	?	

MICRO		
Roost height (m)	Range: ?	
	Niche: ?	
Temperature (°C) **Humidity** (% rh)	Range: 2–8 °C	
	Niche:*	Oct–Nov: 6.2 °C (±2)
		Dec–Feb: 4.5 °C (±2.2)
		Mar–Apr: 4.2 °C (±2)
Roost height (m)	Range: ?	
	Niche:*	Oct–Nov: 91% (±14)
		Dec–Feb: 89% (±13)
		Mar–Apr: 86% (±9)
		Exposed situation: 89% (±15) – 97% (±8)
		Enveloped situation: 81% (±15) – 93% (±9)
Associates	?	

* Wermundsen 2010; Wermundsen & Siivonen 2010.

Table 5.21 Daubenton's bat *Myotis daubentonii* subterranean rock roosting account.

MACRO											
Landforms exploited	Solution cave ✓		Sea cave ?			Mine ✓			Railway tunnel ✓		
Months in which subterranean landforms occupied	Winter		Spring-flux		Pregnant		Nursery		Mating		Autumn-flux	
	Jan ✓	Feb ✓	Mar ✓	Apr ✓	May ✓	Jun ✓	Jul ?	Aug ✓	Sep ✓	Oct ✓	Nov ✓	Dec ✓

MESO	
Roost situation	Threshold ✓	Dark-zone ✓

Roost exposure	Roost position exposure Exposed / Recessed / Enveloped % of roosts (0–100) Values taken from Bezem *et al.* (1964), Bogdanowicz & Urbańczyk (1983) and Daan & Wichers (1968)

MICRO		
Roost height (m)	Range: 0–7.1 m	
	Niche: 3.2–3.6 m	
Temperature (°C)	Range: 0.2–10.4 °C	
	Niche:	Oct:* 9.2 °C / Nov:* 8.9 °C → Oct–Nov:** 6.5 °C (±2.6)
		Dec:* 7 °C / Jan:* 6.8 °C / Feb:* 6.5 °C → Dec–Feb:** 4 °C (±1.6)
		Apr:* 6.4 °C → Mar–Apr:** 3.7 °C (±1.8)
Humidity (% rh)	Range: 71–100%	
	Niche:**	Oct–Nov: 93% (±16)
		Dec–Feb: 91% (±9)
		Mar–Apr: 87% (±12)
		Exposed situation: 92% (±23) – 96% (±10)
		Enveloped situation: 86% (±21) – 94% (±16)
Associates	Natterer's bat *Myotis nattereri*	

* Daan & Wichers 1968.

** Wermundsen 2010; Wermundsen & Siivonen 2010.

Table 5.22 Whiskered bat *Myotis mystacinus* subterranean rock roosting account.

MACRO										
Landforms exploited	Solution cave ✓		Sea cave ?			Mine ✓			Railway tunnel ✓	
Months in which subterranean landforms occupied	Winter		Spring-flux		Pregnant		Nursery	Mating		Autumn-flux	

	Jan ✓	Feb ✓	Mar ✓	Apr ✓	May ✓	Jun ?	Jul ?	Aug ?	Sep ?	Oct ✓	Nov ✓	Dec ✓

MESO		
Roost situation	Threshold ✓	Dark-zone ✓

Roost exposure	

Values taken from Bezem *et al.* (1964), Bogdanowicz & Urbańczyk (1983) and Daan & Wichers (1968)

MICRO		
Roost height (m)	Range: ?	
	Niche: 2.8–3.2 m	
Temperature (°C)	Range: 0–10.5 °C	
	Niche:	Oct:* 10.1 °C
		Nov:* 8.9 °C
		Oct–Nov:** 6.2 °C (±2)
		Dec:* 6.6 °C
		Jan:* 5.9 °C
		Feb:* 6.2 °C
		Dec–Feb:** 4.5 °C (±2.2)
		Apr:* 6.2 °C
		Mar–Apr:** 4.2 °C (±2)
Humidity (% rh)	Range: ?	
	Niche:**	Oct–Nov: 91% (±14)
		Dec–Feb: 89% (±13)
		Mar–Apr: 86% (±9)
		Exposed situation: 89% (±15) – 97% (±8%)
		Enveloped situation: 81% (±15) – 93% (±9)
Associates	?	

* Daan & Wichers 1968.

** Wermundsen 2010; Wermundsen & Siivonen 2010.

Table 5.23 Natterer's bat *Myotis nattereri* subterranean rock roosting account.

MACRO											
Landforms exploited	Solution cave ✓		Sea cave ?			Mine ✓			Railway tunnel ✓		
Months in which subterranean landforms occupied	Winter		Spring-flux		Pregnant		Nursery		Mating		Autumn-flux	

	Jan ✓	Feb ✓	Mar ✓	Apr ✓	May ?	Jun ?	Jul ?	Aug ?	Sep ?	Oct ?	Nov ?	Dec ✓

MESO		
Roost situation	Threshold ✓	Dark-zone ✓

Roost exposure	

Values taken from Bezem *et al.* (1964), Bogdanowicz & Urbańczyk (1983) and Daan & Wichers (1968)

MICRO	
Roost height (m)	Range: 0.2–6.9 m
	Niche: 3.2–3.8 m
Temperature (°C)	Range: –2.1–10 °C
	Niche:* — Oct: ?
	Nov: ?
	Dec: 6.4 °C
	Jan: 6.8 °C
	Feb: 5.9 °C
	Apr: 5.6 °C
Humidity (% rh)	Range: 73–100% rh
	Niche:** — Oct–Apr: 94%
	Exposed situation: ?
	Enveloped situation: 88% (±6) – 95% (±6)
Associates	Daubenton's bat *Myotis daubentonii*

* Daan & Wichers 1968.

** Bogdanowicz 1983; Bogdanowicz & Urbańczyk 1983; Wermundsen 2010.

N.B. Insufficient data were available to produce an account for Nathusius' pipistrelle *Pipistrellus nathusii*.

Table 5.24 Common pipistrelle *Pipistrellus pipistrellus* subterranean rock roosting account.

MACRO											
Landforms exploited	Solution cave ✓		Sea cave ?		Mine ✓				Railway tunnel ✓		
Months in which subterranean landforms occupied	Winter		Spring-flux		Pregnant		Nursery		Mating	Autumn-flux	
	Jan ?	Feb ?	Mar ?	Apr ?	May ?	Jun ?	Jul ?	Aug ?	Sep ?	Oct ?	Nov ?	Dec ?
MESO											
Roost situation	Threshold ✓					Dark-zone ?					
Roost exposure	?										
MICRO											
Roost height (m)	Range: ?										
	Niche: ?										
Temperature (°C)	Range: ?										
	Niche: ?										
Humidity (% rh)	Range: ?										
	Niche: ?										
Associates	?										

N.B. Insufficient data were available to produce an account for the soprano pipistrelle *Pipistrellus pygmaeus*.

N.B. Insufficient data were available to produce an account for Leisler's bat *Nyctalus leisleri*.

N.B. Insufficient data were available to produce an account for the noctule *Nyctalus nyctalus*.

Table 5.25 Brown long-eared bat *Plecotus auritus* subterranean rock roosting account.

MACRO												
Landforms exploited	Solution cave ✓			Sea cave ?			Mine ✓			Railway tunnel ?		
Months in which subterranean landforms occupied	Winter		Spring-flux		Pregnant		Nursery		Mating		Autumn-flux	
	Jan ✓	Feb ✓	Mar ✓	Apr ✓	May ?	Jun ?	Jul ?	Aug ?	Sep ?	Oct ✓	Nov ✓	Dec ✓
MESO												
Roost situation	Threshold ✓						Dark-zone ✓					

Roost exposure

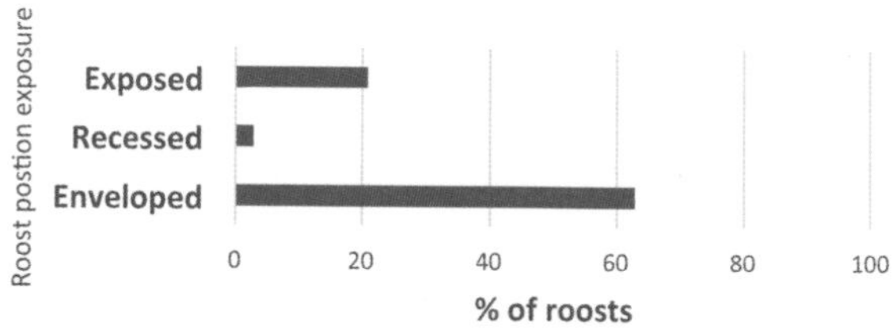

Values taken from Bezem *et al.* (1964) and Bogdanowicz & Urbańczyk (1983)

MICRO			
Roost height (m)	Range: 0.5–6 m		
	Niche: 2.6–5 m		
Temperature (°C)	Range: −3–11 °C		
	Niche:*/**	—	Oct–Nov:** 5.8 °C (±2.8)
		Dec: 5.2 °C	Dec–Feb:** 3.2 °C (±2)
		Jan: 0 °C	
		Feb: 5.3 °C	
		—	Mar–Apr:** 3.4 °C (±1.2)
Humidity (% rh)	Range: 54–100%		
	Niche:**	Oct–Nov: 92% (±19)	
		Dec–Feb: 91% (±11)	
		Mar–Apr: 84% (±9)	
		Exposed situation: 86% (±20) – 94% (±12)	
		Enveloped situation: 85% (±20) – 91% (±11)	
Associates	Barbastelle *Barbastella barbastellus*		

* Daan & Wichers 1968.

** Wermundsen 2010; Wermundsen & Siivonen 2010.

Table 5.26 Grey long-eared bat *Plecotus austriacus* subterranean rock roosting account.

MACRO											
Landforms exploited	Solution cave ✓		Sea cave ?		Mine ✓			Railway tunnel ?			
Months in which subterranean landforms occupied	Winter		Spring-flux		Pregnant		Nursery		Mating		Autumn-flux	

	Jan ✓	Feb ✓	Mar ?	Apr ?	May ?	Jun ?	Jul ?	Aug ?	Sep ?	Oct ?	Nov ?	Dec ?

MESO	
Roost situation	Threshold ✓ / Dark-zone ?

Roost exposure	

Values taken from Bogdanowicz & Urbańczyk (1983)

MICRO	
Roost height (m)	Range: 1.5–2.4 m
	Niche: ?
Temperature (°C)	Range: −0.5–6 °C
	Niche: ?
Humidity (% rh)	Range: 69–90%
	Niche: ?
Associates	?

Table 5.27 Lesser horseshoe bat *Rhinolophus hipposideros* subterranean rock roosting account.

MACRO											
Landforms exploited	Solution cave ✓		Sea cave ✓			Mine ✓			Railway tunnel ✓		
Months in which subterranean landforms occupied	Winter		Spring-flux		Pregnant		Nursery		Mating		Autumn-flux	
	Jan ✓	Feb ✓	Mar ✓	Apr ✓	May ✓	Jun ✓	Jul ✓	Aug ✓	Sep ✓	Oct ✓	Nov ✓	Dec ✓

MESO	
Roost situation	Threshold ✓ / Dark-zone ✓
Roost exposure	

Values taken from Bezem *et al.* (1964) and Daan & Wichers (1968)

MICRO	
Roost height (m)	Range: ?
	Niche: 2–2.5 m
Temperature (°C)	Range: 2–14 °C
	Niche:* — Oct: 10.4 °C
	Nov: 9.7 °C
	Dec: 8.2 °C
	Jan: 8.3 °C
	Feb: 7.9 °C
	Apr: 7.5 °C
Humidity (% rh)	Range: ?
	Niche: ?
Associates	Cave spider *Meta menardi*

* Daan & Wichers 1968.

Table 5.28 Greater horseshoe bat *Rhinolophus ferrumequinum* subterranean rock roosting account.

MACRO											
Landforms exploited	Solution cave ✓		Sea cave ✓		Mine ✓			Railway tunnel ✓			
Months in which subterranean landforms occupied	Winter		Spring-flux		Pregnant		Nursery		Mating	Autumn-flux		
	Jan ✓	Feb ✓	Mar ✓	Apr ✓	May ✓	Jun ✓	Jul ✓	Aug ?	Sep ?	Oct ✓	Nov ✓	Dec ✓
MESO												
Roost situation	Threshold ✓*					Dark-zone ✓						
Roost exposure	?											
MICRO												
Roost height (m)	Range: ?											
	Niche: ?											
Temperature (°C)	Range: 3.5–12.5 °C											
	Niche:**	Oct: 10.5 (±1.1) °C										
		Nov: 9.8 (±1.3) °C										
		Dec: 8.4 (±1.4) °C										
		Jan: 7.6 (±1.6) °C										
		Feb: 6.6 (±1.6) °C										
		Mar: 7.4 (±1.5) °C										
		Apr: 7.5 (±1.2) °C										
Humidity (% rh)	Range: 76–100%											
	Niche: >96%											
Associates	Cave spider *Meta menardi*											

* BRHK Subterranean Rock Database.

** Ransome 1968.

Advice for anyone proposing to survey a rock landform or monitor an artificial landform to deliver a new roost for a specific bat species

Henry Andrews

In this chapter	
Introduction	What this chapter will do; What this chapter will not do
Legislative considerations	The *Wildlife & Countryside Act 1981 (& as amended)*; The *Conservation of Habitats and Species Regulations 2017*
Broad fieldwork considerations	Health & Safety for the bats; Health & Safety for ecologists; Mapping roles and responsibilities; Fieldwork practicalities
Being realistic about what can be achieved	Close-inspection efficacy; Remote-observation of rock face and loose rock landforms; Radio-tracking
Monitoring and surveillance – *they are not the same thing*	Communication; Monitoring design process; Setting the monitoring objective; Agreeing the alarm trigger thresholds; Designing the surveillance prescription; Reporting
Environmental surveillance as part of monitoring	Context; Setting the objective; Surveillance values that will trigger the alarm; Surveillance duration; Surveillance interval; Assessing the data against the alarm triggers
Dataloggers	What dataloggers are and what they do; Examples of two types of datalogger that I use – their strengths and weaknesses; Deploying the loggers; Downloading the data; Replacing the battery; Maintaining the loggers
Historical misapplication of someone else's work in an attempt to place a value on a specific landform	A cautionary tale, with a call to arms …

6.1 Introduction

The accounts of the physical characteristics of different roost situations, as well as the environmental data set out in the preceding chapters, are factually correct and have a practical application. They are, however, incomplete and certainly not all that can be achieved with the advances in equipment that have taken place since many of the studies were published, such as: laser distance measures, digital cameras, hand-held anemometers and hygrometers, dataloggers, lux meters and sound meters, home computers (and with them Excel spreadsheets), and the Internet (and with it social media and file-sharing).

In any case, they do not span all the rocks and rock landforms available to bats in different locations across the British Isles, and much of the data was collected on the Continent. The locations in which the data were recorded have different climates and even different day lengths.

In the early 2020s all these things are within the grasp of every professional ecologist. There is therefore an opportunity to upgrade the knowledge available in the public domain for the benefit of the 17 species of bats known to exploit rock landforms.[46]

To fill the knowledge gaps, standardised survey and recording protocols have been identified for each of the three broad landform groups, and we now have the facility for an online database at www.batrockhabitatkey.co.uk. This chapter offers some informal suggestions for anyone who is now thinking about going out to collect some records.

» **What this chapter will do:** make you aware of obstacles and pitfalls that you may encounter during your surveillance. It will also give you advice on what not to do based on real mistakes made during years of surveys, and some suggestions of things you might like to try in order to complete the forms and interpret what your results might mean. You do not have to follow the advice straight away; if you prefer you can go your own way and make the same mistakes for yourselves.

» **What this chapter will not do:** set out a prescriptive set of one-size-fits-all rules that ignore the fact that: **a)** I have never seen or set foot in the landform you are about to survey, nor do I know you personally or have any idea of what your strengths are; **b)** new scientific discoveries may already have made this book out of date; and, **c)** technological advances are now so rapid that new 'game-changing' equipment comes onto the market every year.

This is **NOT** a set of rules or instructions. It does not include anything that claims to be 'best practice' or even 'good practice', it is just friendly advice and answers to frequently asked questions.

Notwithstanding, this book has engaged with the relevant stakeholders, i.e. the bats themselves. Everything in it is therefore dictated by evidence in the form of data.

Data is coded intelligence: if we decode the message the bats are sending, we will ultimately get so good at knowing where they are, and when they are there, we will no longer be chasing them, but ambushing them – in a good way obviously…

This chapter has some advice in boxes like this one. These boxes hold methods that have been applied to overcome problems. The methods are personal to me, Henry Andrews. I have tried already sharing some of the methods with other people on social media and been met with pomposity and arrogance. If you wish to have a snide dig at the suggestions, you can reach me by putting an upturned galvanised bucket on your head and speaking into it loudly.

6.2 Legislative considerations

Before we progress any further, it is necessary to set out the legislative context of any invasive bat survey. All bat species are protected under two pieces of legislation, comprising:

1. The *Wildlife & Countryside Act 1981 (& as amended)*;
2. The *Conservation of Habitats and Species Regulations 2017*.

6.2.1 The *Wildlife & Countryside Act 1981 (& as amended)*

All bat species are listed under Schedule 5 of the *Wildlife & Countryside Act 1981 (& as amended)* and receive legal protection under Part 1, Section 9, Subsection (4)(b & c) which states:

> Subject to the provisions of this Part, a person is guilty of an offence if intentionally or recklessly —
>
> (b) he disturbs any such animal while it is occupying a structure or place which it uses for shelter or protection; or
> (c) he obstructs access to any structure or place which any such animal uses for shelter or protection.

6.2.2 The *Conservation of Habitats and Species Regulations 2017*

All bat species are listed under Schedule 2 of the *Conservation of Habitats and Species Regulations 2017* (updated by the *Conservation of Habitats and Species (Amendment) (EU Exit) Regulations 2019*). Part 3, regulation 43, paragraph (1) of the Regulations states:

> A person who—
>
> (a) deliberately captures, injures or kills any wild animal of a European protected species,
> (b) deliberately disturbs wild animals of any such species,
> (c) deliberately takes or destroys the eggs of such an animal, or
> (d) damages or destroys a breeding site or resting place of such an animal,
>
> is guilty of an offence.

Note: The offence of damaging or destroying a breeding site or resting place does not include the word 'deliberately', but is an absolute offence that does not require any fault elements to be proved to establish guilt.

Part 3, regulation 43, paragraph (2) states that disturbance of animals includes any disturbance which is likely:

(a) to impair their ability —
    (i)   to survive, to breed or reproduce, or to rear or nurture their young, or
    (ii) in the case of animals of a hibernating or migratory species, to hibernate or migrate; or
(b) to affect significantly the local distribution or abundance of the species to which they belong.

Therefore:

» Where there is evidence to suggest bats are present (such as historical records) the inspection should only be performed by a surveyor who is licensed to disturb bats;[47]

» Where there is no evidence to suggest bats may be present, an unlicensed surveyor might reasonably look for them, but if they are rewarded with a positive result they should withdraw immediately and seek assistance from a licensed surveyor.[48]

6.3 Broad fieldwork considerations

6.3.1 Health and safety for the bats

In the context of a purely visual inspection: **a)** shining a light into a darkened Potential Roost Feature (PRF) is intrusive and will disturb any bat inside; **b)** all endoscope inspections are intrusive and disturbing, as well potentially damaging to the PRF; and, **c)** entry onto a rock face, or loose rock, as well as entry into a subterranean situation by a human, is an intrusion.

All artificial light is disturbing to nocturnal animals. In subterranean landforms headlamps should be used to navigate. Cree torchlight should be used carefully. You do not want blinding torch beams crossing like light-sabres at a Sith convention. There is nothing more frustrating than a bat taking wing and if you have lots of torches going here-there-and-everywhere that is exactly what will happen (and no you aren't an exception, it absolutely WILL happen). The most sensible advice I can give you when mapping is to search slowly and calmly.

This is my approach to photography when I am mapping in subterranean landforms. I tried to get sensible advice and got nothing, so I went and I made mistakes and things went wrong – I disturbed bats and they flew away – it really happened to me and it is really going to happen to you too. Things are going to go wrong, but the bats will come back, and you will learn.

When I find a bat or a group of bats: **1)** I move the torch beam calmly away and stop where I am; **2)** I take the camera out and prepare it fully for the shot I am about to take; **3)** I hand the camera to my runner (I do not hand them the torch because they do not know where the bats are and I do; my runner can follow my torch beam with the camera); **4)** I use a Lenser torch and I turn the beam to the lower power; **5)** I put the beam to the floor and my runner lets the camera adjust to that light level (Mark Latham taught me this trick and it is really helpful); and, **6)** I bring the beam slowly back up to the bat(s) keeping the beam in camera shot so the exposure retains the balance (my runner can then take the photograph and tell me if they need more light). Job done and bats least disturbed.

The same applies to a situation where I need tripods – I do not hold the torch beam on the bat(s) any longer than I need to, or in the time it takes to put the tripods in place the bats WILL have flown and I am left wondering if they were ever there at all.

Even doing this some bats may still take flight, but not as many as might have otherwise. However, if a bat takes wing, continue with the mapping calmly as you would have anyway. Individual bats are not in any risk of physical harm by your presence. Providing you do not try to follow it with the torch but simply ignore it, now that it is awake it will find its own way to another roost position. Obviously if another bat joins it you have to be sensible and think about whether you are causing significant disturbance and abandon the mapping if there is any chance your presence is causing distress.

N.B. *Caving Practice & Equipment* (Judson 1984) has a useful section on subterranean photography and there is also a short section in *The Complete Caving Manual* (Sparrow 2009) that is worth a look.

In repeat surveillance trainees should be very closely supervised with each individual trainee shadowed by a more experienced surveyor taking one-on-one responsibility. However, in any of these situations I ask myself: *Quo bono?* Surveillance should not ever be exploited as a 'bucket-list experience'.[49] If the bats do not benefit from the presence of observers, might there be any additional detrimental effect? I guard against this by keeping numbers to the absolute minimum to achieve the objective.

> I am always firm with any 'newbies' as follows: **a)** they should be seen and not heard; **b)** they should sit until called forward to look; and, **c)** no, they may not take photographs because they always take an age getting the camera ready and then muck about trying to get that perfect shot and this means the bat is in the light too long.

Where endoscopes are used, apply the *Bat Tree Habitat Key (BTHK)* project guidance as follows:

1. Do not risk the endoscope coming into contact with a bat;
2. Withdraw immediately if babies or dependent young are present;
3. Where babies or dependent young are not present, the bat(s) should be in the light for no more than ten seconds;
4. The inspection should be complete within 30 seconds.

No. 1 and No. 2 are **RULES**:

> **Do not EVER risk the endoscope coming into contact with a bat and withdraw immediately if babies or dependent young are present.**

No. 3 and No. 4 are guidance and will depend upon circumstances. However, common sense should prevail and consideration given to how agitated any bat present is getting. The simple test of trial by social media can be helpful: ask yourself, 'would I want anyone seeing how this inspection is going?' If the answer to that question is 'no', then STOP/ regroup/come back with a new approach that mitigates the potential for distress and damage to an acceptable level, or abandon the operation entirely.

Once the search for bats has been performed, the presence of any bats can be taken into account in order that the environmental sampling has the least risk of a disturbance effect.

> **Note:** Be careful when attempting to use a 17 mm lens because the head is so much wider than the snake it can get stuck. It is better to use a snake with a lens-head that is as close to the diameter of the snake as possible.

In the context of a physical search that will require the movement of individual rocks, you must first consider the fact that to move a rock you are going to be lifting weight and applying weight to the rock surface around the weight you will lift (even if it is only your own body weight). Think about where the load-bearing weight will be applied to the rock face. If there is any risk of squashing anything beneath the surface or damaging the landform, stop and mitigate or do not search.

Next, think about the movement of the weight. Prising-up weight results in a rolling pressure which will move the rocks upon which the weight rests; this is unacceptable. The weight must be raised in a single movement that will decrease all the pressure it exerts over the entirety of the surface at once. The weight cannot roll, hinge or bump the surface but must be lifted like a Jedi lifts an X-wing.

Now, think about where you are going to put the weight you have raised while you search underneath it. If you cannot put it down carefully somewhere it will not result in damage, do not search.

Finally, think about how you are going to return the landform to be exactly the same as it was before you searched it. If you cannot: STOP, and do not search it.

Regardless of the search context, a valuable exercise is to set out why the surveillance is being performed as though it was an explanation that would be presented to the bats themselves in order to gain their permission.

6.3.2 Health and safety for ecologists

It is up to the individual survey team to ensure they have the correct qualifications, experience, access equipment, personal protective equipment, insurance and risk assessment, and that their equipment is regularly assessed and certified as safe to use.

You are referred to:

» **Mitchell-Jones, A. & McLeish, A. 2004.** *Bat Workers' Manual,* **3rd edn. Joint Nature Conservation Committee, Peterborough**
  − Chapter 2 is dedicated to Health and safety in bat work.
» **Sparrow, A. 2009.** *The Complete Caving Manual.* **Crowood Press, Wiltshire**
  − This has 20 Golden Rules for Safety as well as an entire chapter dedicated to first aid, survival and rescue. Although the book is directed at cavers the recommendations can be equally applied as a minimum in every rock survey situation (and in fact a survey of anything anywhere).

6.3.3 Mapping roles and responsibilities

This section could easily have been discussed under health and safety because it has such a bearing on minimising injuries and maintaining morale. It is recommended that mapping is performed by at least three people, whose roles and responsibilities are as follows:

» **Searcher** − The searcher does the search. They are responsible for nothing more than finding the PRF and the bats in them. They should be prepared to lay on the ground in mud, on a cave floor, and move around in cold, damp and filthy conditions.
» **Runner** − The runner passes the searcher equipment and takes it from them. They are responsible for all the equipment. Their job is that of a theatre nurse: when the searcher says 'endoscope' they hand the endoscope, and when the searcher says 'torch' they take the endoscope and pass the torch. Their role is to keep the searcher safe, so they should be taking in the wider picture of the surroundings: pointing out any PRF entrance they can see from their position away from the rock, as well as obstacles and ensuring that the ground around is adequately lit.
» **Scribe** − The scribe is responsible for writing the recordings down. They are essentially a shadowy slug that moves through the landscape slightly away from the searcher and runner, and stopping at a convenient place to sit and take down the values. In subterranean situations they should maintain sufficient distance to not be under foot but still hear every spoken value. The scribe should be organised and

thorough. They are ultimately responsible for ensuring every value is recorded accurately and the forms are stored as cleanly as possible. They do not deal with equipment because that will result in their hands getting damp and dirty, and that will ruin the forms.

Decide on a leader. Leaders: watch your two teammates carefully and rotate roles every half an hour to give people a chance to rest, and make sure that the scribe roll does not result in someone getting cold. When roles are defined for that period, stick to them; runners should not be looking for bats but watching the searcher to ensure they do not stray into danger or miss anything.

Each person has two hands, but they will need both to negotiate obstacles, so think about how you will transport equipment and keep the records dry and safe. A Daren jar in a round-base bag is a must for keeping completed recording forms dry and can be clipped to a belt for easy transportation. A nice new, clean and shiny Daren, bag and clip are illustrated in Figure 6.1.

Figure 6.1 A round-base bag, Daren jar and belt clip.

6.3.4 Fieldwork practicalities

These facts are relevant regardless of which landform you are proposing to survey.

» **Fact 1** – You are NEVER going to be on flat ground: the base of a face will be littered with boulders; scree only occurs on steep slopes; and, aside of railway tunnels, every other subterranean landform has an uneven floor. Think carefully about the terrain because it has a bearing on every part of the survey. At the very least, you will need non-slip footwear and both hands free.

> I find walking boots to be about the worst footwear possible: the soles are inflexible, they are not waterproof (I do not care what the manufacturer says – they are NOT waterproof), the laces come undone, and the eyelets catch on absolutely everything. The best footwear is cheapo Wellington boots. These have the best grip on bare rock. This is the reason cavers wear them. When I'm surveying rock, I find myself spending the greater amount of my time moving slowly and clumsily across uneven ground like a chimp, and not yomping over wide distances. When I wore walking boots, I got wet and cold feet and I slipped (you are not special: you WILL slip too). If you have to crawl, the eyelets will snag and then snap and your boot will be a big heavy and expensive flip-flop.

» **Fact 2** – The ground is always going to be at least damp, and more often than not really wet. It will be slippery and covered in sharp projections that will cause damage if you fall over onto them. It will get recording forms and all your equipment wet and muddy. Think carefully about how you will move around to survey, and how you will record on paper in damp conditions.

> Whenever I am working with rock faces, or entering a subterranean landform, I wear: **1)** a climbing helmet (not a builder's hard-hat because this does not have a chin-strap and will fall off if I have to clamber about to look in things); **2)** a caving suit or hardwearing overalls; **3)** a caving belt (to clip things to and for someone to grab me if I need them to pull me up and out of anything); **4)** kneepads; and, **5)** general purpose builders' gloves.
>
> I wear this lot because I am going to bring my head into direct proximity of sharp rocks, and spend as much time kneeling, sitting, bent over and lying down, as I do standing upright.
>
> Recording in the wet is a problem that has no easy answer. I do not take electrical 'tablets' into the field because they inevitably go wrong (maybe yours does not, but I have seen enough people lose their temper in the field with a malfunctioning tablet to be convinced that paper is still the best option). It is easiest to record in a weather-writer, but I keep the clean forms in a separate container and move them five at a time into the recording weather-writer so that I do not end up with a block of papier mâché. I also take a small towel with me to dry my hands.

» **Fact 3** – You cannot put anything down, because it will roll and slip and then fall down a crack. If you manage to retrieve it, it will be soaked and covered in mud. You do not therefore want your recording equipment on your back – you want it on your front. If you use a rucksack you will put it down, it will fall over, and all your equipment will spill out down the slope. If you put anything down it will roll and get lost. You therefore need to think carefully how you will carry all your torches, measures and meter, etc. while you move between PRF.

> In surface situations a Barbour coat is excellent because they have deep pockets and do not tear on sharp rocks. Alternatively, I have 58 Pattern 5-piece canvas webbing which: **a)** has deep pockets for my torches, measures, meters, etc.; **b)** spreads the weight over me; **c)** is indestructible; and, **d)** is comfortable when I am clambering about. In addition, I can assign specific kit to specific pockets/pouches so I am not constantly looking for things and, as the pouches are big enough to rummage about in, I do not need to put anything down.

» **Fact 4** – Recording on your own is miserable. Think about how you will maintain your morale in the face of adversity.

> Teams of three are great, if you are going with more people, ideally you will increment them in groups of three. This means that you can allocate roles, and one person can stay in the most comfortable situation where they stay clean and dry and can keep the recording form clean and dry while they fill it in. You can then have two people searching while one scribes. Then change roles. When I am leading a trip, I make sure I have something for my teammates to nibble on: Haribo, Maoam, Kendal Mint Cake, whatever … it really does help.

» **Fact 5** – If you are not disciplined you will achieve nothing. Set a realistic goal that does not depend on finding a bat, and stick to it.

> When I find a PRF I stop and assess whether it is worth recording. If it is worth recording, I do not move on until I have recorded it. The more I move around, the more chance I have of slipping and damaging myself. Having done a huge amount of tree survey and no small amount of rock surveys, I do not try inspecting the whole outcrop/cave/mine, etc. in the hope of finding a bat and then go back to record the best empty PRF because: **a)** I will inevitably keep my torch in one hand and I know I need both hands empty to be able to grab on if I slip. So, I move, inspect, put equipment back in pockets, move, inspect, put equipment back in pockets, etc.; and, **b)** what I think is the best PRF may not be what a bat thinks is the best PRF.
>
> Before I became disciplined I had quite a few demoralising days when I returned home feeling confused and unhappy. Now that I am disciplined, I always get records. These records might be of empty PRF and not the roost that will make my name, but I do always have a map and some tangible data to enter. And the next time I visit that site, I can simply inspect the PRF without having to get all the pens, and forms, and tape measures out all over again. Essentially, even on a day with no bats, I finish with a satisfying sense of progress.

» **Fact 6** – People have told me a lot of twaddle about kit. They really have. I now take this approach: if I meet someone with three different models of the same thing and they give me a florid review of each model, they have three different things that are not quite right. When I meet someone with one, and it is really old: it works and that is the one that I want.

A brief comparison of equipment items that I have trialled as part of this project, with an appraisal of their reliability, is provided in Table 6.1.

Table 6.1 Equipment that is consistently reliable and that which is not.

TASK	RELIABLE	UNRELIABLE OR DOES NOT WORK
Being able to see in the dark	Petzl, Fenix (Scurion if you are millionaire – you can wear it round the house and offset the expense by saving on your electricity bill)	Cree headlamps that claim they are waterproof and then go to strobe before switching off – avoid. Anything with a movement sensor in the lamp – avoid
Being able to see into PRF	LED Lenser torches – you will only buy one because they are reliable and nigh-on indestructible (I have not had good experiences with the head torches though)	Multifunction cree torches – they have every function you do **not** want and never the one you do. They randomly scroll to strobe just as you find a bat, and then switch off completely. The battery charger may burn your house down – avoid
Measuring heights	A cheap laser distance measure or a clinometer	An expensive laser rangefinder that relies on GPS
Measuring distances	A laser distance measure, trundle-wheel or fabric tape measure (the advantage of a tape measure in a cave system is that it can be laid down over the floor and then the search performed knowing when to stop, you then move the tape again and work in incremental 30 m units, which means you can stop anytime knowing exactly how far in you have mapped	A steel tape measure, fine for small lengths such as PRF entrances, but when used for any measurement beyond comfortable arm-span: **a)** it makes a tremendous amount of noise across all frequencies; **b)** it is impossible to manipulate above your head; **c)** it is amazing how often that little steel tip gets jammed in some tiny crevice, which means that; **d)** you snap the measure and leave part of it hanging as a testament to your stupidity
Taking short-distance photographs	A camera phone	A digital camera: the lens is too far in from the edge so it is difficult to get it to where you want it
Viewing and photographing a cavern ceiling	A torch on one tripod and a camera with a zoom lens on another	Anything other than a torch on one tripod and a camera with a zoom lens – the photograph will be blurred
Looking in PRF that are above your head	Putting the endoscope snake and extension through a couple of sections of a carp fishing pole: lightweight and easy to manipulate (a brilliant idea from Dr Danielle Linton, for which she should be paid due tribute)	—

6.4 Being realistic about what can be achieved

This section was written to highlight that the collection of 'definitive' data in a single surveillance campaign is not a realistic expectation, and certainly not when applying an untested, concept-driven framework – that is, as opposed to a data-driven framework. This book is working towards a data-driven framework, but it is a long way off the definition of any quantifiable confidence level. The fact that every rock situation is different means that every survey will be different too, yet one thing is obvious: there is no 'one-size-fits-all' strategy.

6.4.1 Close-inspection efficacy

Obviously, it would be great if people did climb about like Spiderman but it is accepted that (just as with the BTHK project) in the greater proportion of rock face surveys the inspections will be from the ground up to around 2 m, and potentially up to 5.5 m if a surveyor's ladder is used (with appropriate training and a Risk Assessment, obviously). At the moment, there is so little data that all data is good data!

Notwithstanding, the inspections will encounter a high proportion of features that extend over 1 m into the face and/or around right-angle bends. While it is sometimes possible to manipulate the tip of an endoscope round such a corner by pre-bending the tip and then turning it like a key, the potential to injure a bat (or another organism) that might be round the corner should be anticipated.

As a result of the problems associated with the accessibility of PRF in unstable faces, and the complexity of the internal structure of rock face PRF, overall survey coverage may be unsatisfactory and there will be a lot of inconclusive investigations of the PRF that can be reached but cannot be comprehensively searched. This is inevitable, and anyone performing a consultancy survey of a rock face should be prepared to use remote-observation to assess the status of rock landforms.

In surveys of loose rock, unless you are following a radio-tagged bat the probability you will find a bat using close-inspection is significantly low. In any case, in most situations close-inspection of loose rock will be impractical if not irresponsible. Even where the bat is tagged, although you may be able to identify the location of the tagged individual, you will not know whether there are other bats in other locations, which means that as you move to find the tagged bat you risk killing another bat, and also killing yourself by starting a landslide.

And so we come at last to the subterranean landforms, and to give some idea of the efficacy of the close-inspection method here, we refer to the extraordinary work of Punt & van Nieuwenhoven (1957). These guys caught all the bats in one study cave and attached radioactive rings to them, they then used a Geiger counter to detect the locations in which the bats were roosting. Their results demonstrated that 20–45% of the bats present were in enveloped situations. Another study in the Netherlands found even worse odds, in that twice as many bats were recorded in enveloped situations as were recorded in situations where they would be easy to see, and that most bats roosted individually; so, lots of individuals peppered through the system in hard-to-find PRF (Daan 1973).

In most consultancy surveys I perform, rock landforms require a combination of close-inspection and remote-observation.

6.4.2 Remote-observation of rock face and loose rock landforms

The issue with emergence surveys of rock landforms is one of scale. If you know where the PRF is, and you have a clearly defined entrance, then the methods you would use for a building or tree are equally applicable. If you are assessing the status of a large-scale surface in which there are a multitude of openings then the situation is VERY different.

If you are attempting to assess the status of a quarry face then you will have the added constraint of safe stand-off distances.

We could spend a lot of time trying to anticipate every scenario but that would stifle discussion and innovation.

The advice I will share is that whatever approach you choose, do a 'dry-run' in the day. Literally, take all your kit out onto the site and set it up in the daytime to see how and where you will deploy equipment at night, and then write a method statement setting out EXACTLY who is to do what, where and when. All the problems of uneven ground and damp conditions are multiplied at night.

Thermal Imaging really comes into its own with the rock landforms, but if you are going to use thermal imaging and you are not used to the kit, go out and perform the survey in the middle of the day when birds are moving around. This will give you an idea of effective range in that site, and also allow you to fiddle around with the various colour options. I favour setting it to black-and-white, with cold to black, that way the sky is not blinding and does not overpower the surface that is being surveyed. Remember that the image you are seeing on the screen isn't real time, it is a sequence of recalibrated images that refresh periodically. Check what the specifications say in the manual about overcoming this so you do not have gaps in the recording: you do not want to be looking for a solution to a 'screen-freeze' in the middle of the survey.

Keep in mind that although the pistol-grip models (such as the FLIR E75) are superb, they also weigh a lot and you are going to be holding that thing up for up to two hours. I find emergence surveys of rock faces using thermal imaging very tiring on my arms, wrists and also on my eyes.

Finally, electrical kit does not like water. Rock landforms often occur in rainy places and even on a dry summer evening, they attract humidity. Think about how you are going to ensure the kit stays dry and functional. Have an emergency strategy for an unexpected rain shower: it is going to happen to you and you do not want to be running through boulders to find somewhere dry to put the camera (not much point in drying it after you have smashed it on a rock).

I am well known for my dislike of dawn-return surveys: swarming is fine, but dawn-return is bloody boring, bloody cold and can be bloody dangerous with people driving to and from surveys dog-tired. Who in consultancy just does bat work? I do not; I have plants, newts, reptiles, birds and every other mammal to survey, and I cannot write off days of good weather to prepare and recover from a dawn-return. Who lives in a totally quiet house where they can sleep during the day? Where is the evidence that dawn-return is so damn important??? There is a stack of evidence that demonstrates that evening emergence is best (I wrote about it at length in *Bat Roosts in Trees*), yet those luddites keep banging-on about dawn-return. Well, here is my company policy:

> **Emergence surveys every time people – safety and sanity come first and always.**

When someone finally falls asleep at the wheel and kills a young family on the school run perhaps this idiocy will stop.

6.4.3 Radio-tracking

With all the constraints to close-inspection and remote-observation we might be tempted to step back and hand over to the experts and their radio-tracking wizardry, but … the scientific literature is peppered with accounts of the frustration of attempting to use radio-tracking to identify roost positions in hibernacula. While the autumn-flux may prove successful, on the whole the winter appears to be fraught with complications. Two reasons given are the much faster drop in battery power resulting from cooler temperatures, and the loss of the bat due to them roosting deeper into the landform. Kristiansen (2018) provides an account of a simple test of the efficacy of a transmitter in scree, which demonstrated that the signal was attenuated/blocked when the unit was behind rocks and did not reach further than 30 m and sometimes as little as 15 m.

6.5 Monitoring and surveillance – they are not the same thing

6.5.1 Communication

The Internet has allowed ecologists to communicate with each other across disciplines and around the globe. This means that ecological consultants and academics can now share knowledge and use that knowledge to improve our approaches to solve problems.

BUT that only works if we speak the same language and are confident that the words we use have a specific meaning. One word is incorrectly used more than any other: 'monitoring'.

In the vast majority of its use in published accounts, the word 'monitoring' is used when in fact what has been performed is 'surveillance'. For example, by far the greater proportion of 'monitoring' that has been performed following the implementation of a European Protected Species Licence (EPSL) across the length and breadth of the UK has not been monitoring at all: it has been surveillance and as no one has been collating the data to any particular end, it is a total waste of resources; time and money down the drain.[50]

Monitoring, as defined by JNCC (1998), comprises intermittent surveillance carried out in order to ascertain the extent of compliance with a **predetermined standard**.

Monitoring is therefore the comparison of surveillance data against a predetermined trigger for action. Used correctly, monitoring will ALWAYS be paired with a list of specific management measures that are triggered when the surveillance shows that the situation is outside a desired limit.

In simple terms, monitoring is used to **move** or **maintain**. When moving, we have a positive threshold: our goal. This goal is the definition of success. When maintaining, we have a negative threshold below which management intervention will be required to return the situation to a point within the desired limit. If we do not have a trigger for alarm, we are not monitoring, what we are doing is surveillance.

For example, in a scheme to deliver wood pasture, the surveillance might be a count of the trees each year and the annual measurement of the diameter at breast height (dbh) of trees in the same month each year.

The triggers might be: **a)** a situation where over 10% of the trees were dead or moribund (which would require beating-up to replace losses); and, **b)** a dbh of 20 cm (at which point the tree is pollarded for the first time).

Recorded diligently these data would allow managers to forecast the survival rate of initial planting and first pollarding, and the likely management expense in each year when setting out to create another new wood pasture.

Surveillance informs monitoring and it can also be performed in isolation simply as a chronicle. However, using the word monitoring when you mean surveillance makes you look like an idiot to the Police, the armed forces, and everyone who works in medicine and engineering.

6.5.2 Monitoring design process

In an ecological context, monitoring is often concerned with ensuring that the management for maintaining the quality of a given site is effective (Hellawell 1991). A functional monitoring programme is simple and typically comprises four parts: 1) an objective; 2) an alarm with a definite trigger; 3) a surveillance prescription; 4) a format for reporting.

6.5.3 Setting the monitoring objective

The first stage in the design of a monitoring scheme will be the definition of the objective: the specific goal of the project (Margoluis & Salafsky 1998; Hellawell 1991).

In the context of this project, the objective will almost always be the delivery and maintenance of an environment within the realised niche of the target species, and as close to the fundamental niche as is possible. The objective will therefore be informed by the proven and documented niche of the target species. As a result, it may be that surveillance is required in order to identify what that niche is!

As, in the most basic terms, the objective will be the minimum acceptable state or condition, it is therefore vital that the objective is clearly defined at the outset with due consideration given to qualitative attributes as well as quantitative ones (Legg & Nagy 2006; Hill *et al.* 2005; Salafsky *et al.* 2002; Kleinman *et al.* 2000; Vos *et al.* 2000; JNCC 1998; Margoluis & Salafsky 1998).

It is all too easy to concentrate disproportionally on a target species within monitoring. This is what happens with EPSL monitoring: we go, we look for the species, it is not there, we report, we do not know why it is not there and there is no penalty when it is not there. The licence is satisfied, the box is ticked, the species has lost out and we have learned nothing.

Outside the context of a veterinary surgery, you cannot monitor faunal individuals. Surveillance that only counts those individuals has no practical use whatsoever. As the individual animals have free will, and can move in response to factors beyond the control of the monitoring, the failure of surveillance to detect them in any one year (even if this were to accurately represent absence) is of no value in determining whether the environment was favourable to support them. The point is that the roost might have been suitable or unsuitable and unless we monitor the environment we will not know. Even if all aspects of the roost environment are within the fundamental niche range, the bats might choose to roost somewhere else: you cannot force them to occupy it so you cannot monitor them because unless you start a captive breeding programme and lock them in the roost, you cannot manage their population.

This is important – you can control a population by culling, but maintaining a population requires control of recruitment and you are not going to achieve that with one roost.

As the target species will only persist within an area if the environment is suitable, it is far more important to focus all the resources on the successful delivery of that environment: this we can control and therefore monitor.

6.5.4 Agreeing the alarm trigger thresholds

The objective having been established and clearly set out, it is then necessary to select a 'trigger' (the deviation from the desired situation) above or below which an effect is considered unacceptable and action is then triggered. In order that ambiguity does not lead to uncertainty, each alarm should be triggered by recordings taken by a specific meter and compared against a clearly defined fixed scale.

6.5.5 Designing the surveillance prescription

Having determined the alarm trigger threshold, the surveillance 'prescription' may then be designed.

To assess the position of the environment in relation to the niche, the surveillance will record the baseline conditions in an existing roost (Tucker *et al.* 2005). In the context of this project, you would map and then record the environment (e.g. habitat over the entrance, types of PRF, air movement, temperature range and fluctuation, humidity range and fluctuation, illuminance). This will tell you what the species is currently occupying in your site.

As monitoring schemes should be focused entirely upon the objective (Tucker *et al.* 2005), the prescription should be designed with the sole purpose of triggering the alarm and focus on attributes that are objectively *measurable* (Cottle & John 2009).[51] The surveillance prescription will therefore be environment-specific (Cottle & John 2009), and use methods that are easily replicated year on year in order to be repeatable over long periods of time, by different surveyors with potentially wide ranges in skills and experience.

Once defined, every effort should be made to ensure that all aspects of the prescribed method (i.e. form, frequency, timing, duration and equipment) are repeated as closely as possible.

Ideally monitoring will also contribute to the diagnosis of the causes of change and assess the success of actions taken in order to maintain standards or to reverse undesirable changes (Tucker *et al.* 2005). This is particularly pertinent to untested management practices where there is the potential that the proposed management regime may be flawed and so inform the ongoing refinement of the methods used until a proven prescription for each individual action is achieved (Salafsky *et al.* 2001).

> Whether or not a scheme is functional monitoring or not can be decided by a simple dichotomous question: Will the data gathered: **a)** trigger an alarm if something is going wrong? **b)** inform practical management action; **and, c)** inform the definition of the fundamental niche of the organism? If the answer is yes, you are doing monitoring, if the answer is no, you are doing meaningless box-ticking.

6.5.6 Reporting

Monitoring reports are job-sheets; they are directions for management with a description of why the management is needed and what that management is to achieve. They are not wordy prose but simple instructions. The best monitoring reports have the instructions on page 1 – literally the cover page – and the reason for the management etc. should follow so that a contractor can read them if they are interested, but will not be put off by them and therefore not get the 'message' in the report.

Fundamentally, a good monitoring report is a table with three columns: **1)** what needs doing; **2)** when; and, **3)** who by.

However, if we are to learn, the person writing the report should have a stab at a hypothesis as to what might be the cause of failure to deliver the optimum, with a prediction of what the outcome of the management will be and why. This should be set out in plain language, and based upon the full body of monitoring data gathered thus far.

6.6 Environmental surveillance as part of monitoring

6.6.1 Context

To give this section a meaningful context, it will work to the premise that the surveillance is being performed as part of a monitoring programme in an artificial roost. The roost has been built as part of an enhancement package for a development, and is to benefit a specific bat species. We already know the bat species is present in the immediate locale and we have connected the landform to the network of commuting routes exploited by the species.

6.6.2 Setting the objective

I have found that monitoring is best if the objectives have a series of three realistic goals. These are the small victories that maintain project morale. Before that, we need two definitions, as follows:

» **Realised niche** – The realised niche is what a species can tolerate without abandoning a territory. Consultants often use the term 'suboptimal' to describe situations: well, suboptimal is the species-realised niche – the species can occur here but it is not the ideal. Very little in nature is optimal for any species, and all of it goes through annual fluctuations.

» **Fundamental niche** – The fundamental niche is the optimum. However, it is not a single fixed value and will have a tolerance range either side before it passes out of the fundamental and into the wider realised niche range. Outside the realised niche range it becomes unfavourable and then ultimately hostile.

Objective 1

I start by identifying the realised niche range the species tolerates and the narrower fundamental niche range the species wants within the realised range. Delivery of an environment that is always within the realised range is my first objective and any situation outside this range will trigger the alarm and require remedial action.

Now, although we have tolerance range data from most species and have set it all out in Chapter 5, these data are biased towards Finland, Estonia and Poland. While this does represent best available evidence it is not ideal and, as we do not have the temperatures recorded outside the landform on the days the bats were exploiting specific temperatures inside, it is not possible to make an informed comparison. What we do know is that the average February temperatures are as follows: Finland –22 to –3 °C, Estonia –6 °C, Poland –5 °C, and England +1 to +7 °C. The difference is pretty obvious and leaves us with a nagging doubt as to whether the thresholds are transferable.

Ideally, we will gather **baseline** data of our own in several different roosts that are in the locality we are going to build our artificial roost, so that we can at least get a realised niche tolerance range; this leads me to my first mistake:

» **Mistake 1.1** – I have been seduced by the quick fix and taken 'off-the-shelf' ideas from a wide variety of sources that have not been successful for me. I hear how successful the various designs are for other people and find myself confused and demoralised; why does mine not work?!? I now collect baseline environmental data every time both inside an existing roost landform and immediately outside, and that means that I can see why the design is not working in my situation and fix it, and that leads to better designs.

Objective 2

When I eventually secure Objective 1, my trigger sensitivity is increased by narrowing the parameters to a position where the alarm will sound if the environment is anything outside the fundamental niche; that which represents the optimum situation.

This is the really hard part, because every adjustment I make to improve conditions in one respect, may result in a negative response somewhere else. There are two mistakes I have made at this stage:

» **Mistake 2.1** – The tendency to 'overshoot' the target in an attempt to move the temperature a little one way or another. However, if I overshoot this does at least demonstrate that the fundamental niche is possible, so I add/remove some insulation or increase/decrease the size of an aperture, etc., by 50% and see what the result is.

» **Mistake 2.2** – Being seduced by the potential to deliver one roost that will fit all months, which ends up as one roost that is never offering the fundamental niche in any month, rather that something that is at least stable and valuable in a particular period.

Objective 3

Finally, having moved the conditions towards the fundamental niche, I want to keep them there. This is important when we consider how long the monitoring might have to last. The question is, am I delivering an enhancement that will not require maintenance? If it is all concrete and brick then I might be, but if it is perishable materials then it certainly will not be, and I need to be sensible to that fact and ensure that my monitoring takes it into account so that maintenance is performed to keep the thing functional.

6.6.3 The value of manually recorded values and automatic remote logging

In the next couple of sections, we are going to look at recording kit. This divides into two groups, as follows: manually operated meters, and automatic dataloggers. The two are both vital because they have different strengths.

Manually operated meters include: **a)** rotating vane anemometers for measuring air movement; **b)** therma-hygrometers for measuring temperature and relative humidity; **c)** lux meters for measuring illuminance; and, **d)** sound meters for measuring how noisy a roost is.

Automated dataloggers comprise a battery-powered unit that records the temperature, or both the temperature and relative humidity over long periods of time. The time interval the units take measurements at can be set using the manufacturers software and the units typically last a year on one battery.

The two approaches – manual and automated – can be used individually or in combination. However, when I am getting a baseline what works best for me is making manual metering the primary data collection method, with automated logging subordinate, and when I am monitoring what works best for me is making automated logging the primary data collection method with manual metering subordinate.

Dataloggers are fantastic, but unless I make regular visits at short intervals to see where the bats have moved to, I will not have a reliable baseline (i.e. what the species was experiencing in its original roost) because the bats might have moved to a different position and my logger does not follow them.

I might attempt to inform my baseline by installing loggers in every possible situation the bats might occupy, and pair each logger with a trailcam, but even if I could guarantee that the loggers and cameras would not get stolen, the outlay in terms of manpower and tech would be considerable. For this reason, while loggers are fantastic for monitoring, they are not a panacea, and for establishing a baseline data set of the environment the bats are actually occupying in an existing roost, manual readings give a more reliable picture. It is accepted that this will be an incomplete picture, but the bits that are there are reliable.

With manual readings you can see the bat is there and measure the environment the bat is occupying on that specific day and what the environment is doing outside. You can then compare the manual recording against the baseline weather data recorded on a logger deployed outside. When you have the baseline data you can produce an initial hypothesis of what the realised niche is for that species in your locality.

The air movement, lux and noise environment will also require baseline and tolerance range data (i.e. what the species is known to tolerate across all roosts) for comparison. Unfortunately, an ambitious plan to collect data across multiple species in a wide variety of roosts was spoiled by COVID-19. For example, in the sites which have limited human ingress the risk of giving COVID to the bats meant that we did not enter, and the manned sites were shut, but even if we had been granted access, a good deal of activity that would have been generating draughts, light and noise was reduced and the levels simply would not have been representative of the true situation (e.g. roosts in a quarry were stagnant, and dark and quiet because the staff were all furloughed: doors shut, lights off, machines silent and no traffic in or out).

And so, we come to monitoring. Now, until the bats adopt my artificial roost, if I only take manual readings of the temperature and relative humidity, I will have insufficient information to determine anything other than whether the environment was within suitable limits at the moment I took the reading. However, if I am using dataloggers, the assessment to see whether the roost is delivering the realised niche range is easy: I simply download the data recorded by the loggers and compare it against my baseline data and the relevant species account provided in Chapter 5 to see whether the minimum temperature recorded on the logger remains above the minimum temperature tolerated by the species, and whether the maximum temperature recorded by the logger remains below the maximum temperature tolerated by the species.

For example, Figure 6.2 shows the temperature and humidity data gathered on a logger in an artificial chamber that has been built for greater horseshoe bats. The realised niche range has been marked on the Figure to illustrate whether the environment is correct.

The results are pretty damning: the temperature and humidity are on or below the base range alarm trigger thresholds in most months, and the temperature fluctuations span 7 °C in some months. If I were a bat, I would not occupy a roost in which the temperature went up and down like a yo-yo; it is hardly conducive to a good week's torpor. In any case, I might prospect the feature on a day when the temperature was uncomfortably cold or warm and not bother coming back at all.

And so, we made some initial alterations and the results show up immediately in the logger data, as illustrated in Figures 6.3 and 6.4.

Figures 6.3 and 6.4 suggest that the alterations had stabilising effect on temperature fluctuations, and a wholly positive effect on relative humidity. However, they would be compared against the control data on the temperature logger outside the system to ensure that the improvement was not simply attributable to a milder winter.

> **Note:** The data also suggest the fact that temperature is harder to manipulate than humidity, but that is not always true and unless the artificial landform is built somewhere that is naturally damp there will be long battles with humidity too.

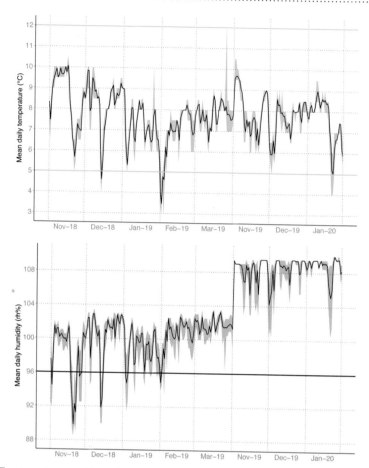

Figure 6.2 Temperature and humidity data gathered on a logger in an artificial chamber that has been built for greater horseshoe bats shown against the realised niche range as described by Ransome (1971).

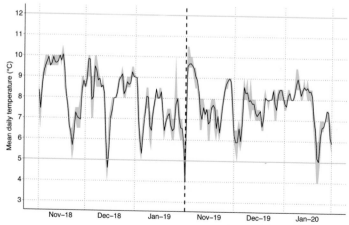

Figure 6.3 The mean daily temperature (°C – black line) and daily fluctuation in temperature (shaded yellow) recorded from November 2018 to January 2019 and November 2019 to January 2020. The solid yellow lines represent the limits of the realised niche range (i.e. what the species will tolerate). The vertical dashed line divides the environment before and after alterations were made.

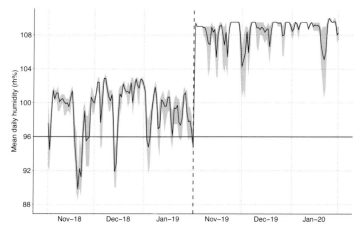

Figure 6.4 The mean daily humidity (rh% – black line) and daily fluctuation in humidity (shaded blue) recorded from November 2018 to March 2019 and November 2019 to January 2020. The solid blue line represents the fundamental niche (i.e. what the species really wants). The vertical dashed line divides the environment before and after alterations were made.

However, in this case the improvements really have had a positive effect and now that the realised niche has been broadly delivered in the meso-environment, we can begin moving the environment into the fundamental niche range in the micro-environment by compartmentalising the wider space and providing smaller features that can be insulated to offer the stable temperature. The loggers will show the response to the environment to each new alteration we make.

However, it is at this stage that Mistake 2.2 can creep in. If we were in an optimistic mood, we might be tempted to overlay the data in Figure 6.3 with the fundamental niche ranges for each month, and that is exactly what has been done in Figure 6.5.

We might think that this one artificial roost is getting close to offering the fundamental niche across multiple months, but that 2018/19 was better than 2019/20. In fact, what the temperature fluctuations illustrate is that the roost is still far too responsive to the weather and climate outside, we are not even close to a useful landform because what the bats need is reliability.

At the moment the roost environment is never stable, and to a bat it means that it is unreliable. Figure 6.6 illustrates what 'reliable' look like in logger data that was recorded in a greater horseshoe bat roost.

These data are real: the logger is not malfunctioning. The roost environment is fixed and has been delivering the same environment day after day, month after month and year after year for millennia.

It would be a rare situation for one chamber in a solution cave to deliver a year-round roost that tracks the outside temperature to offer exactly the right environment in each month. What is more common is for different chambers (even different landforms), to each deliver a slightly different temperature in each month. Effectively, what the logger data shown in Figure 6.5 are showing us is that rather than being close to delivering a universally optimum nirvana, what we actually have is a comprehensively unsuitable environment.

In fact, Objective 2 is to deliver **stability** and **reliability** at the fundamental niche position. Delivering a single landform that offers the fundamental niche across the autumn-flux, winter and spring-flux may therefore need to encompass three different roost features, each offering a different constant temperature, and yes, I know a lot of you already know this but there are a lot of people who do not.

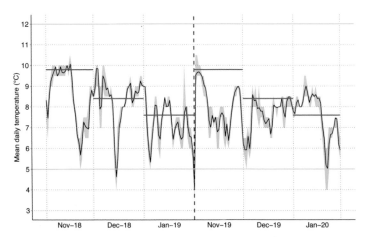

Figure 6.5 The fundamental niche positions occupied by the greater horseshoe bat on the temperature gradient (as described by Ransome 1971) overlaid on data recorded in an artificial roost feature.

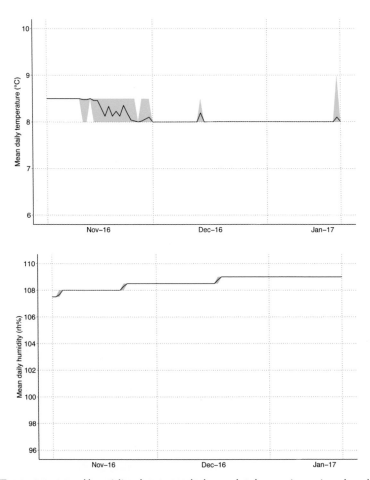

Figure 6.6 Temperature and humidity data recorded on a datalogger in a mine chamber occupied by greater horseshoe bats *Rhinolophus ferrumequinum*.

Note: I have deliberately included a difficult project here. The reader should not be put off trying to create a roost landform because this looks like a lost cause.

First, this is part of a wider project to deliver roost habitat to several different species. So far, we have serotine *Eptesicus serotinus*, Daubenton's bat *Myotis daubentonii*, Natterer's bat *Myotis nattereri* and brown long-eared bats *Plecotus auritus*. We are now working to try to offer five different areas to lesser and greater horseshoe bats and this is proving more difficult but also very interesting. In fact, this book was begun to inform this tricky project, and without the project there might not have been a book at all!

Secondly, more enhancements are seeing more improvements and ultimately this project will be a success: we will deliver the fundamental niche (although whether the local greater horseshoe bat population choose to colonise it in my lifetime is another matter entirely).

There is still more work to be done here … but my loggers will tell me when I can stop.

6.6.4 Environmental surveillance equipment for collecting the values that will trigger the alarm

My triggers will ideally be set to specific thresholds of: **1)** air movement; **2)** temperature; **3)** humidity; **4)** illuminance; and, potentially also **5)** noise. All five of these factors can be measured using meters that are available from nhbs.com. As the recordings can be taken in specific positions, there is no room for subjectivity.

Air movement can be recorded using a hand-held anemometer. The air movement is measured because manipulating it is something that is easy to manage by installing partition walls, curtains of material such as shed felt,[52] or discrete shelter features (i.e. a roost within a roost). The anemometer can also be used to measure the effect of any remedial installation as it is installed, and tailor it to the desired response (e.g. 'a bit more … more … tiny bit more … stop … take it back a bit … there: perfect'). The basic rule is that the more air movement you have the lower the temperature will be and the lower the humidity. I take the measurements at ceiling level using a 'selfie stick' and also at the floor level, at the entrance, middle and back of the roost feature, and obviously in and around any specific roost feature.

Temperature and relative humidity can be recorded manually using a hand-held combination meter, or continuously using data loggers. As has already been discussed, unless you are working in a secure site where a trailcam can be left with the logger, or the visits can be performed sufficiently frequently that the location of the bats in relation to the logger is comprehensively documented, baseline data is most reliably and cheaply done with manual meters.

Some of the manual meters have a probe that can be inserted into the entrance of an enveloping PRF, but be aware that the probes are 2 cm in diameter. Typical examples of an anemometer and manual temperature/humidity meters are shown in Figure 6.7.

However, while they are great for getting an idea of the local baseline, with a hand-held unit all you will get is an individual reading and not the range data that you will need for monitoring. What you really want to trigger your monitoring alarm is a set of temperature and humidity loggers that you can deploy and leave for a year at a time. Section 6.7 goes into datalogger management in detail.

Illuminance can also be recorded using a hand-held lux meter.

Noise is measured using a hand-held sound-level meter and a bat detector. This isn't strictly necessary in all situations, but if you are working in very noisy environments

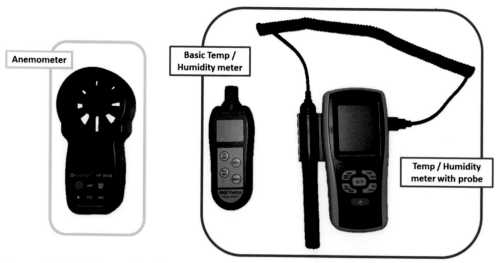

Figure 6.7 Typical examples of an anemometer and manual temperature/humidity meters.

> **VITALLY IMPORTANT NOTE:** The temperature/humidity meter with the probe comes with a little stringy ariel-type thermometer that plugs into the meter on the top – YOU WILL NEED THIS EVERY TIME YOU GO OUT! The probe is great until it encounters supersaturated air, at which point the unit has a tantrum, sounds an alarm, and stops reading anything until the probe dries down again. When this happens, the screen shows the text 'TYPE K' which meant nothing to me. However, what it means is 'plug in the stringy thing and at least the unit will function to record the temperature'. Unless you know what is going on, you will refer to the instructions (if you still have them) and they are unhelpful. Worse still you may have discarded the stringy thing or not have it with you. This results in an epic Rumpelstiltskin stomp session on the meter, which disturbs the bats, pollutes the cave and scares the people who are with you and suddenly aware that they are many metres below ground where no one will hear their screams …

then establishing the dB(A) that different species will tolerate at different frequencies may be helpful in identifying why a particular situation might not be occupied.

Typical examples of a lux and noise meter are shown in Figure 6.8.

6.6.5 Surveillance duration

Monitoring does not have tidy boundaries: it continues until the objective is secured. If the objective will require maintenance, the monitoring will continue forever.

The duration is therefore dependent upon two factors: **1)** the time it takes to deliver the fundamental niche; and, **2)** the need for repairs in order to maintain the fundamental niche.

At the outset there will be a need to assess whether the roost is delivering the realised niche range, i.e. the widest proportion of the environmental gradient that the species can tolerate and still survive. At the very least a full span of values will have to be recorded across the occupancy period. For instance, if the roost is to offer a hibernaculum the baseline of values will span the period November through April. If any remedial action is required, it will not be possible to assess the results of the work until the following

Lux meter

Sound (noise) meter

Figure 6.8 Typical examples of lux and noise meters.

November through April period and if the results are not satisfactory, then further manipulation will be required, and there will be another year before their effect can be assessed.

Once the realised niche is delivered, attention turns to delivering the fundamental niche, i.e. the narrower range that the species really wants on each environmental gradient.

When the fundamental niche is finally delivered, monitoring will have to continue in order to assess the need for repairs.

Regardless, even if by some stroke of luck, the roost immediately delivers the fundamental niche and requires no maintenance whatsoever, it may be several years before it is colonised. To explain, Poulton (2006) investigated the interval between the deployment of artificial bat-roost boxes and the date bats were first recorded roosting within them. With the species-specific results ordered from the shortest interval to the longest, the average interval between deployment and adoption reported is as follows:

» Pipistrellus spp. – 1 year and 87 days;
» Brown long-eared bat *Plecotus auritus* – 1 year and 268 days;
» Noctule *Nyctalus noctula* – 2 years and 36 days;
» Bechstein's bat *Myotis bechsteinii* – 2 years and 110 days;
» Natterer's bat *Myotis nattereri* – 2 years and 214 days;
» Daubenton's bat *Myotis daubentonii* – 3 years and 93 days.

Although the sample sizes were different between the species, the results do represent a reliable interval for undisturbed 'static' situations (i.e. boxes deployed in habitat that had been effectively unchanged for decades if not hundreds of years). It is reasonable to suppose that a long-lived mammal that clearly has a mental map of its territory will be cautious about entering torpor in an entirely new rock landform that is itself in a situation that has been recently landscaped.

6.6.6 Surveillance interval

In the first year I will make two visits to: **a)** deploy the loggers and take the first round of manual air movement, illuminance and noise readings; and, **b)** retrieve the loggers, download the data, replace the battery and redeploy, and take the second round of manual readings for air movement, illuminance and noise.

Whether I shorten the interval thereafter will depend upon the results and what remedial action is required, but until the fundamental niche is delivered it can be predicted that there will be at least two visits annually: **1)** to get the data; and, **2)** to make any alterations needed to move the environment into a stable position within the desired range.

6.7 Dataloggers

6.7.1 What dataloggers are and what they do

Automated dataloggers comprise a battery-powered unit that records the temperature, or both the temperature and relative humidity over long periods of time. The time interval the units take measurements at can be set using the manufacturers software.

6.7.2 Examples of two types of datalogger that I use – their strengths and weaknesses

I favour the Corintech Ltd, EasyLog USB Logger sold by www.nhbs.com. However, I cannot always afford them so I also use the Elitech RC-51H which is sold on www.amazon.com. The two models are illustrated in Figure 6.9.

The first difference is that the EasyLog is dark grey so easy to hide (it's finding them again that's the hard part). They have no buttons and no screen, so there is nothing to worry you in the field. The unit has the vents on the sides and the end capped, so you can deploy it vertically in either direction and also horizontally. Bottom line: once you switch the EasyLog on there is nothing to complicate the deployment other than finding somewhere to put it. The instructions for operation are also simple and require no particular skill in software manipulation.

Another major plus to the EasyLog is that you can download the data using the EL-Datapad, which is shown in Figure 6.10.

The datapad fits in your pocket and all the logger data can simply be transferred onto it, so you do not need to take a laptop down a mine. This is a significant plus if you are

Figure 6.9 Left: the Corintech EasyLog USB. Right: the Elitech RC-51H.

Figure 6.10 The Lascar EL-Datapad.

not using the loggers in rely and taking them back to the office to download. However, the datapad downloads the data as a text file which is an irritation when what you really want is the data in Excel.

In contrast, to the grey of the EasyLog, the Elitech loggers are orange and therefore difficult to hide anywhere outside an ochre pit. However, if you are deploying the logger in an open situation this is easily rectified with some black electrical tape. The unit also has: **a)** a screen with hieroglyphics to send you worrying coded messages; **b)** a really irritating little on/off button that you have to avoid putting any pressure on while you deploy it so you do not accidentally switch it off; and, **c)** the vents opening out of one end, which is also annoying when you consider that deploying it vertically means that it fills with water in the first 24 hours and registers 108% humidity for the entire year. Other than that, it's great and the power button can be easily protected by wrapping the logger in bubble-wrap, which also makes deployment and retrieval easier because you can stuff it into a suitable crevice without having to hunt for one that is exactly the right size. In fact, regardless of which logger you chose, bubble-wrap is a must for deployment because it makes the logger squishy and much easier to deploy.

6.7.3 Setting up the logger for deployment

Both sorts of logger come preset to record at 15-minute intervals and the way the loggers are programmed to record the date and time value is not user-friendly. If you are looking at the initial suitability of an environment all you need is one value for each 24-hour period; midnight is a good starting point. Unfortunately, you cannot programme the loggers to record two discrete times, so if you want midnight and midday you will have to pair two loggers. If you are looking at fluctuations in a 24-hour period you only really need every hour, but recognise that this will mean you have to format 720 entries for every 30 days deployment. If the logger is deployed for a year, you will have 8,760 values to reformat because the date and time are one long value stream in a single cell an example of which looks like this: 2021-01-22 13:24:59. Before the data can be manipulated for analysis in Excel, you will have to separate the day, month, year and time into four columns, and 8,760 of those are going to be pretty laborious.

EasyLog

To change the time on the EasyLog all you have to do is plug it into your laptop and do the following:

1. Download the EasyLog software from their website;
2. Insert the logger into the USB port on your laptop;
3. Open the 'EasyLog USB software' from your desktop;
4. Click 'Set up and start the USB data logger';
5. Ensure frequency is set to 24 hours or 1 hour;
6. Do not set temp or humidity alarm (these are preset to off anyway);
7. Delay the start until the time you want the logger to record the first reading.

N.B. The EasyLog set-up is very straightforward, but I cannot find a way to assess the condition of the battery without taking it out and using a battery tester. Personally, I find this irritating.

Elitech

The Elitech software is more involved than the EasyLog, but it takes less than an hour of playing with to get the hang of it. Even if you are a luddite, the process to reprogramme the recording interval is easily achieved by finding a young person who has the patience and good enough sight to read the instructions. For the benefit of the more mature reader, I had a young person create a streamlined set of instructions for me and they go with Figure 6.11 as follows:

1. Download the Elitech Log software free from their website;
2. Plug the logger into your laptop;
3. Open the software;
4. Go to 'Parameters' which is in the bar at the top of the screen;
5. Look for 'Log interval' which is at the top of the control panel;
6. Set H = 1; M = 0; S = 0 to record every hour;
7. Then using the button on the bottom left save the parameter.

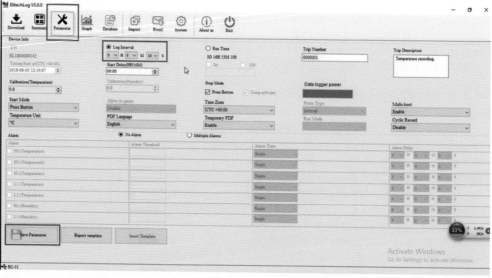

Figure 6.11 The operating interface for setting the datalogger recording interval.

N.B. You can also name the logger in the Trip number and description boxes at the top right of the control panel. Note also that the battery power level is shown as a big green bar which is really helpful.

6.7.4 Deploying the loggers

Dataloggers are wonderful pieces of kit, if you accept that they are far from perfect. The following text includes all the other irritations that they bring.

You will need at least two loggers: one to measure the regime in the landform and another to measure the regime outside for comparison across years. It is not good saying you have the conditions correct and walking off site, if you have just had a good weather year and next year is going to return the conditions to a point outside the realised niche. You therefore see how the temperature outside affects the temperature inside and this allows you to predict what the temperature inside was likely to have been in previous years.

N.B. The logger that is deployed outside can just be a temperature logger, which is cheaper.

Where more PRF are present, more dataloggers can only add to the power of the analysis. Ultimately, use as many dataloggers as it takes to get a representative description of the environments offered within the cave/mine both in roost positions and the wider system.

And now for Sod's laws:

» **Law 1** – The EasyLog has to be started before you leave the office, but as the Elitech has a button on it I have been tempted to wait. However, if I do not start the Elitech logger while I am in the office, mine have a nasty habit of not switching on in the field.

  Always start the dataloggers before you leave the office: they are absolutely going to malfunction if you wait until you are in the field – this is a fact of life.

» **Law 2** – The bats always roost in situations where it is difficult to deploy the logger. You will always be uncertain as to whether the deployment situation is representative of the roost environment. You are therefore absolutely going to struggle to deploy the logger in the field. In open situations the only way you could get the logger up might be to use a piton, or drill a hole in the wall to put a bolt in. Neither solution is applicable to solution caves where to piton or drill would be an act of vandalism. Clearly you cannot put a logger in an enveloping PRF if to do so would exclude the bat. However, none of this matters, because the moment you leave the site the logger will be stolen or eaten by an Orc.

I try to get the logger into a representative situation that is as close to the height of the roost above the floor as possible.

6.7.5 Downloading the data

Regardless of which logger you use, you will have to download the relevant software from its respective website. If you are taking a laptop out into the field and you do not have the software, you are going to have a wasted day. Download the software as soon as the logger arrives, or do it now; you know what do now anyway. Seriously, put this book down and go and download both software apps!

If you are downloading the EasyLog straight to a laptop the process is easy and you are unlikely to get into trouble, BUT if you want an Excel file (and you do) you will have to download it onto a laptop and manually select the download format. With the Elitech there is the potential to lose everything if you let it automatically download and then exit the software without forcing it to download in Excel!!!

EasyLog download

If you are using the EL-Datapad to download an EasyLog the instructions are as follows:

1. Ensure the Datapad is fully charged using cable provided;
2. Take the Datapad to the logger;
3. Turn on the Datapad (press and hold the power button);
4. Plug the logger into the Datapad;
5. Click 'Stop Logger & Download';
6. Click 'Next' (this should download the data and say 'data saved');
7. Click 'Done';
8. To redeploy click 'Set-Up Logger';
9. Then 'Quick Set-Up';
10. Then 'Immediate Start' (this should tell you to remove the logger which can be redeployed with clear plastic cover tightly secured).

Transferring the data from the Datapad to a laptop is also simple, as follows:

1. Make the folder you want to put the data into on your laptop;
2. Connect the hard-drive and Datapad to the PC using the cables provided, and then open the files;
3. Drag and drop the text files from the Datapad into the folder on your laptop;
4. Eject the hardware.

Elitech download

When you plug the Elitech RC-51H into your laptop it will automatically generate a PDF report. When it has finished, **DO NOT EXIT – IGNORE ITS DOWNLOAD LEAVING THE POP-UP BOX OPEN, AND THEN:** open up the Elitech log software (you have to use the software if you want to do anything more than download the automatic PDF). Go to *Graph* on the top toolbar, then when you see the data appear click *Export Data* on the bottom left and you'll see an option to select *XLS*. Select this and you will get the data in a spreadsheet that you can actually use – and you want to be able to use it.

6.7.6 Replacing the logger battery

The battery will need replacing every year. To do this you will need:

1. A small flat-bladed screwdriver with a sharp blade;
2. Some sticking plasters for when you stab the hand holding the logger with the screwdriver;
3. A new battery;
4. A laptop to plug the logger into to ensure it is now functioning and has the correct date and time;
5. Another new battery for when the first one doesn't work and just results in an error message on the datapad/laptop that means nothing to you;
6. A new logger for when the second battery doesn't work either.

I recommend doing the battery changes in the office and using loggers in relay: one in, one out. That way you are not pressed for time and if the unit malfunctions you are in a warm and well-lit environment where you are a lot less likely to go into meltdown.

6.7.7 Maintaining the loggers

The logger manufacturers recommend calibrating the loggers every year. You have to send them away to be calibrated and prices are around £50–75. Obviously, I do not recalibrate my loggers: if they appear to be saying anything weird, I toss them and buy a new one. Other than that, I keep them running because regardless of calibration, the sensors have a lifespan of about 4 years. So, after three years I put a new one next to the old one, that way if the new one dies or is stolen, I may still have something useful on the old one.

I think I have now covered everything I might usefully tell you, and so (in the tradition of the Habitat Key projects) we close with something that will hopefully join some unconscious dots and get everyone thinking about the bigger picture of niche theory.

6.7.8 Looking at the data

Figures 6.2 to 6.6 were created using 'R' statistical analysis software but if you are not familiar with that software Excel can be used to create line charts to see what is happening to the environment. Examples of simple temperature investigations using Excel are provided in Figures 6.12 and 6.13.

For those people who want to play with their data, I recommend *Managing Data Using Excel* by Dr Mark Gardener: it is a no-nonsense approach to getting the most out of Excel and has useful exercises to practise specific operations – a really worthwhile investment.

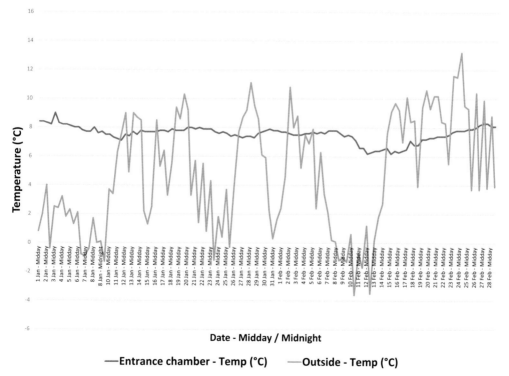

Figure 6.12 The relationship between the temperature inside a chamber occupied by Daubenton's bat *Myotis daubentonii* and lesser horseshoe bat *Rhinolophus hipposideros*, and the outside temperature.

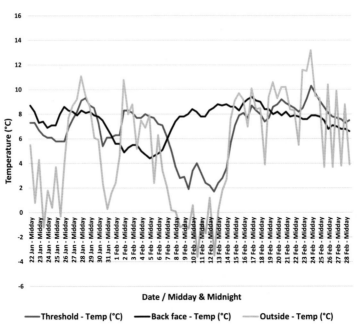

Figure 6.13 The relationship between the temperature at the threshold and on the back face of a short passage occupied by a lesser horseshoe bat *Rhinolophus hipposideros* and the outside temperature.

6.8 Historical misapplication of someone else's work in an attempt to place a value on a specific landform

This last section uses two pieces of scientific work as examples. They are fantastic pieces of science and have delivered valuable data.

» **Fact 1** – The historical misapplication of these data by me and others is in no way a reflection of their value; it is a reflection of how easy it is to misapply good data and draw the wrong conclusion

Historically, there have been attempts to use scientific accounts of subterranean environments to construct a valuation criterion for the assessment of subterranean landforms. These criteria have generalised across all species, as though the landforms were not all as individual as a fingerprint and 'bats' were some multi-headed winged caterpillar and not individual species with very different needs.

» **Fact 2** – I am guilty of attempting this myself when I first came into ecological consultancy and before I got to grips with niche theory.

The misuse of studies fall into two broad camps, as follows:

Camp 1 – The 'size matters and bigger is better' school

This applies the rationale that bigger landforms have more species and are therefore better. An example of this would be the misuse of long-term surveillance studies. Figure 6.14 is an illustration of how the misunderstanding is achieved.

The data demonstrate that in the surveillance as the systems get larger, so the bat fauna becomes more speciose. The data are fantastic but they were not intended for a practical application as a valuation tool. When I looked at these data in conservation terms, the two stand-out species were obviously barbastelle and serotine. My initial perception was that barbastelle and serotine only appeared when a system was greater

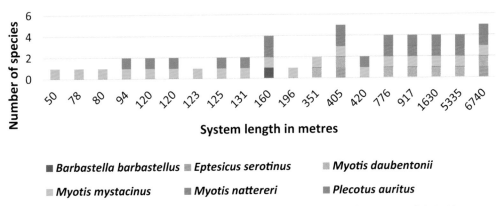

Figure 6.14 Long-term data gathered in Poland and written up by Piksa & Nowak (2013).

than 150 m and 400 m in length respectively, BUT I now know these guys favour the portal and threshold of the system, so their presence is likely to have nothing to do with the length of the system.

The take home message is to look at what the data in the paper were gathered for. Piksa & Nowak (2013) is a herculean piece of work, but it does not include the distances the bats were roosting from the entrance, because that was not a part of the study. The data therefore cannot be used as a valuation tool. Remember that every individual bat species has a specific niche and you cannot generalise across species.

Camp 2 – There is one universal environment that all bat species favour above all others

The most widely known classification is that described by Ransome (1968), which recognises three basic subterranean environment types, comprising:

1. **Constant-temperature (CT) systems** – Typified by a cave with only one entrance and a more-or-less level floor. Such caves have no airflow and temperatures vary little throughout. Ransome (1968) reported a stable mean temperature of 9.5 °C. As well as solution cave examples, sea caves, mine adits and blocked railway tunnels may also offer a constant-temperature environment. Although they may not be biodiverse in terms of the number of species, they may nevertheless offer the particular environmental niche favoured by an individual species and therefore hold high numbers of that species (e.g. Daubenton's bats (Kokurewicz 2004)).

2. **Variable-temperature (VT) systems** – Typified by a cave with more than one entrance and/or vertical open shafts. Such systems have a variety of different conditions over their length and, in Ransome's 1968 study, offered temperatures ranging from about 3 to 14 °C due to airflow caused by two or more entrances. This airflow offers a range of temperatures both above and below mean temperature for the region, but the systems also had static temperature conditions in certain parts. As well as solution caves, mines offer examples of variable-temperature systems.

3. **Variable-low-temperature (VLT) systems** – Typified by a cave with one entrance but a descending floor which results in cold air sinking down into the sloping entrance and pushing warm air out over the roof. Such systems are typically cooler and only offer temperatures below mean climatic levels outside. Notwithstanding, even variable-low-temperature systems may exhibit different temperatures over their length.

This is the most widely misused classification. My own misinterpretation came about because Dr Ransome (1968) observed that constant-temperature systems showed the least stable number of greater horseshoe bats over the period October through May. In contrast, variable-temperature systems supported the most stable numbers, and individual caves in which the highest numbers of bats were recorded on a single date were also invariably variable-temperature systems.

I simplified the situation to assume that the least stable the numbers were, the less valuable the landform is. But this overlooks three facts: **1)** the paper is not saying that one sort of cave is bad and another is good, it is simply providing a division of the landforms into three for future work to see when the bats occupy each of the different types; **2)** not all bats are greater horseshoe bats; and, **3)** bats are not primary hibernators and, unlike dormice and bears, bats are flexible in their use of torpor and seek different conditions at different times.

Having begun to realise my own mistake (and then had some awkward discussions with various other consultants), I went back to the data in Dr Ransome's 1968 paper and did some initial calculations in my own rudimentary way. I then had my conclusions checked by Dr Mark Gardener by a process of statistical interrogation. The findings are presented below specifically for ecological consultants in a series of six facts.

1. Dr Ransome's study was limited to greater horseshoe bats alone: none of the results can be extrapolated to any other bat species.

2. All three types of system were exploited by greater horseshoe bats: they therefore **ALL** have value to greater horseshoe bats.

3. The difference between CT and VLT systems is not statistically significant. What is significant is that CT and VLT have fewer entrances than VT systems. What that means is that a VT system may encompass all three systems in association with each entrance, and all under one roof but in differing extents, i.e. they may have only one or more situations that offer a CT and others that offer a VLT, but these may be different distances from the entrances and different sizes, and therefore more-or-less attractive to greater horseshoe bats at different times, and able to accommodate different numbers. If a temperature below or above that outside the cave is what is required in a particular week, and the extent of that environment offered within a specific VT is too little to accommodate all the bats present, then there will be emigration to a simple CT or VLT system outside the VT. And without those systems the bats might be in real trouble.

4. The CT/VLT systems may only be used for short periods, but they are vital to support the bats that occupy the VT systems: the population needs all the systems to maintain their numbers. Furthermore, on a given day occupancy of CT and VLT systems may outweigh the occupancy of the VT, and the data do not suggest that this is predictable to specific months.

5. The passage length does not appear related to whether it is CT, VT or VLT: you can have a big CT and a small VT. This feeds back into the earlier Camp 1 discussion.

6. The only significant difference in the total count was between VT and CT, with VT having significantly more bats. When the counts in CT and VLT are combined 50% of the overall count of bats was recorded in them in some months.

I have subsequently written to Dr Ransome to own up to my 'valuation' misunderstanding and he was clear that he had not intended his work to be used as a valuation tool for consultants. Neither did he claim that the drawings covered the complete range of such structures. The descriptions were intended to illustrate the structural features that affect airflow – or the lack of it – through any underground system. Dr Ransome also pointed out that his 1968 paper uses data recorded over a single winter that happened to be 1962/3, the winter of 'The Big Freeze', when snow lay from late December to mid-March. This resulted in a total lack of data in January 1963, and uneven sampling in that year, which inevitably had a bearing on the statistical analyses.

My parting messages are:

» When you are collating historical accounts, get the original paper and if possible, contact the author to ensure you are not making any silly mistakes.

» When you get the paper, read the data in the tabulation and figures carefully and put it into Excel so you can use it to try to scope-in and not scope-out.

» Never generalise across bat species: they have evolved to be different species that avoid conflict by occupying different niches on each environmental gradient, BUT those niche positions are not static, they are like a combination lock that changes across the seasons.

» Record baseline **environment** data before you destroy a roost and use that data as the alarm thresholds for the new roost.

» Remember that ecology is a science and not a religious cult: focus on the environmental values that you can record with a meter and not some subjective untestable artificial concept index that relies on belief rather than proof.

» Data is coded intelligence and without it we learn nothing. Every survey that does not end up in Excel is an opportunity missed.

» Where evidence is lacking, get off your bottom, put on your wellies and go and get it: the bats will tell us what they need, we just need to learn to listen to them.

» Bats do read books, but only if they have written them …

Solution caves, sea caves, mines and railway tunnels that are notified as Sites of Special Scientific Interest (SSSI) for bats

Table A1 Solution caves that are notified as SSSI in respect of bat species.

ROCK	SSSI	TYPE	M. bra	M. dau	M. mys	M. nat	P. pip	P. aur	R. hip	R. fer
Limestone	Alyn Valley Woods and Alyn Gorge Caves	Solution cave							✓	
Limestone	Axbridge Hill and Fry's Hill	Solution cave							✓	✓
Limestone	Banwell Caves	Solution cave								✓
Limestone	Blackcliff-Wyndcliff	Solution cave							✓	
Limestone	Buckfastleigh Caves	Solution cave							✓	✓
Limestone	Burrington Combe	Solution cave				✓			✓	✓
Limestone	Cernydd Carmel	Solution cave				✓				✓
Limestone	Cheddar Complex	Solution cave		✓	✓	✓		✓	✓	✓
Limestone	Chudleigh Caves and Woods	Solution cave								✓
Limestone	Coedydd ac Ogofâu Elwy a Meirchion	Solution cave				✓	✓	✓		✓
Limestone	Coedydd Capel Dyddgen	Solution cave								
Limestone	Coed y Gopa	Solution cave and mine		✓		✓			✓	
Limestone	Coed y Mwstwr Woodlands	Solution cave							✓	
Limestone	Coedydd Parkmill a Cwm Llethrid	Solution cave			✓	✓			✓	✓

ROCK	SSSI	TYPE	M. bra	M. dau	M. mys	M. nat	P. pip	P. aur	R. hip	R. fer
			BAT SPECIES							
Limestone	Craig Adwy-Wynt a Choed Eyarth House a Chil-Y-Groeslwyd	Solution cave and mine							✓	
Limestone	Creuddyn	Solution cave and mine							✓	
Limestone	Crook Peak to Shute Shelve Hill	Solution cave							✓	
Limestone	Cwm Clydach	Solution cave							✓	
Limestone	Ebbor Gorge	Solution cave							✓	✓
Limestone	Ffynnon Beuno and Cae Gwyn Caves	Solution cave				✓		✓		
Limestone	Gower Coast Rhossili to Port Enyon	Solution cave								✓
Limestone	Graig Fawr	Solution cave							✓	
Limestone	Little Hoyle & Hoyle's Mouth Caves & Woodlands	Solution cave				✓			✓	✓
Limestone	Llanddulas Limestone and Gwrych Castle Wood	Solution cave and mine		✓		✓	✓	✓	✓	
Limestone	Lydstep Head to Tenby Burrows	Solution cave							✓	✓
Limestone	Masson Hill	Solution cave and mine	✓	✓	✓	✓		✓		
Limestone	Moel Hiraddug a Bryn Gop	Solution cave								
Limestone	Mynydd Llangatwg	Solution cave	✓	✓	✓	✓		✓	✓	✓
Limestone	Napps Cave	Solution cave							✓	✓
Limestone	Penrice Stables and Underhill Cottage	Solution cave							✓	
Limestone	Pierce, Alcove and Piercefield Woods	Solution cave							✓	
Limestone	Poole's Cavern and Grin Low Wood	Solution cave		✓	✓	✓				
Limestone	Potters Wood	Solution cave								✓
Limestone	Ruabon/Llantysilio Mountains and Minera	Solution cave and mine							✓	
Limestone	Siambre Ddu	Solution cave							✓	
Limestone	St Dunstan's Well Catchment	Solution cave				✓				
Limestone	Torbryan Caves	Solution cave							✓	✓

ROCK	SSSI	TYPE	BAT SPECIES							
			M. bra	M. dau	M. mys	M. nat	P. pip	P. aur	R. hip	R. fer
Limestone	Twyni Chwitffordd, Morfa Landimôr a Bae Brychdwn	Solution cave							✓	✓
Limestone	Upper Wye Gorge	Solution cave							✓	✓
Limestone	Via Gellia Woodlands	Solution cave and mine	✓	✓	✓	✓		✓		
Limestone	Wookey Hole	Solution cave								✓
Sandstone	Castle Hill Deer Park and Windy Pits	Cave	✓	✓	✓	✓		✓		
Sandstone	Gowerdale Windy Pits/Peak Scar	Cave		✓	✓	✓		✓		
Sandstone	Foxwood	Cave							✓	

Table A2 Sea caves that are notified as SSSI in respect of bat species.

ROCK	SSSI	TYPE	BAT SPECIES		
			E. ser	R. hip	R. fer
Limestone	Arfordir Penrhyn Angle	Sea cave		✓	✓
Limestone	Berry Head to Sharkham Point	Sea cave		✓	✓
Limestone	Castlemartin Range	Sea cave		✓	✓
Limestone	Creigiau Rhiwledyn	Sea cave		✓	
Limestone	Pen y Gogarth	Sea cave		✓	
Limestone	Pwll-du Head and Bishopston Valley	Sea cave			✓
Limestone	Stackpole	Sea cave	✓		✓

Table A3   Mines that are notified as SSSI in respect of bat species.

ROCK	SSSI	TYPE	B. bar	M. myo	M. bec	M. bra	M. dau	M. mys	M. nat	P. pip	P. aur	R. hip	R. fer
Basalt	Mwyngloddia Wnion a Eglwys Sant Marc	Mine											
Chalk	Eaton Chalk Pit	Mine										✓	
Chalk	Glen Chalk Caves	Mine	✓			✓	✓	✓	✓	✓	✓	✓	
Chalk	Grime's Graves	Mine					✓	✓					
Chalk	Hangman's Wood & Deneholes	Mine					✓	✓	✓		✓		
Chalk	Horringer Court Caves	Mine	✓			✓	✓	✓	✓		✓		
Chalk	Little Blakenham Pit	Mine				✓	✓	✓	✓		✓		
Chalk	Mole Gap to Reigate Escarpment	Mine			✓		✓	✓	✓	✓	✓		
Chalk	Shide Quarry	Mine							✓				
Chalk	Stanford Training Area	Mine	✓			✓	✓	✓	✓		✓		
Granite	Cabilla Manor Wood	Mine				✓	✓				✓	✓	✓
Limestone	Appleby Fells	Mine				✓		✓					
Limestone	Beer Quarry and Caves	Mine			✓	✓	✓	✓	✓		✓	✓	✓
Limestone	Belle Vue Quarry	Mine		✓	✓								✓
Limestone	Box Farm Meadows	Mine											✓
Limestone	Box Mine	Mine				✓							✓
Limestone	Brown's Folly	Mine			✓						✓		✓
Limestone	Buckshraft Mine & Bradley Hill Railway Tunnel	Mine and Railway tunnel				✓					✓	✓	✓
Limestone	Chilmark Quarries	Mine			✓	✓	✓		✓		✓	✓	✓
Limestone	Coed y Gopa	Mine and solution cave				✓	✓	✓	✓			✓	✓
Limestone	Combe Down and Bathampton Down Mines	Mine				✓	✓	✓	✓		✓	✓	✓

ROCK	SSSI	TYPE	B. bar	M. myo	M. bec	M. bra	M. dau	M. mys	M. nat	P. pip	P. aur	R. hip	R. fer
Limestone	Compton Martin Ochre Mines	Mine											✓
Limestone	Cotswold Commons and Beechwoods	Mine										✓	✓
Limestone	Craig Adwy-Wynt a Choed Eyarth House	Mine and solution cave										✓	
Limestone	Creuddyn	Mine and solution cave										✓	✓
Limestone	Devil's Chapel Scowles	Mine	✓									✓	✓
Limestone	Fonthill Grottoes	Mine			✓		✓	✓	✓		✓	✓	✓
Limestone	King's Wood and Urchin Wood	Mine											
Limestone	Llanddulas Limestone and Gwrych Castle Wood	Mine and solution cave					✓		✓		✓		✓
Limestone	Masson Hill	Mine and solution cave				✓	✓	✓					✓
Limestone	Minchinhampton Common	Mine										✓	
Limestone	Old Bow and Old Ham Mines	Mine				✓	✓	✓			✓	✓	✓
Limestone	Ruabon/Llantysilio Mountains and Minera	Mine and solution cave										✓	
Limestone	Townsend	Mine											✓
Limestone	Upper Wye Gorge	Cave										✓	✓
Limestone	Via Gellia Woodlands	Mine and solution cave				✓	✓	✓	✓		✓		✓
Limestone	Westbury Brook Ironstone Mine	Mine										✓	✓
Limestone	Wigpool Ironstone Mine	Mine				✓		✓	✓			✓	✓
Limestone	Winsley Mines	Mine				✓						✓	✓
Mudstone	Cadair Idris	Mine										✓	
Mudstone	Ganllwyd	Mine					✓					✓	
Mudstone	Garth-Eryr	Mine							✓		✓	✓	
Mudstone	Gwynfynydd	Mine					✓					✓	
Mudstone	Mwyngloddiau Llanfrothen	Mine										✓	

ROCK	SSSI	TYPE	B. bar	M. myo	M. bec	M. bra	M. dau	M. mys	M. nat	P. pip	P. aur	R. hip	R. fer
Mudstone	Phoenix United Mine	Mine											✓
Quartzite	Chwarel Cambrian	Mine				✓	✓	✓			✓	✓	
Sandstone	Aberdunant	Mine							✓			✓	
Sandstone	Allt y Main Mine	Mine										✓	
Sandstone	Coedydd Beddgelert a Cheunant Aberglaslyn	Mine										✓	
Sandstone	Hembury Woods	Mine										✓	
Sandstone	Morcombelake	Mine										✓	
Sandstone	Mwyngloddfa Cwmystwyth	Mine					✓		✓		✓		✓
Sandstone	Mwyngloddfa Mynydd-bach	Mine										✓	
Sandstone	The Stiperstones & The Hollies	Mine										✓	
Sandstone	Westerham Mines	Mine				✓	✓	✓	✓		✓		
Shale	Haytor and Smallacombe Iron Mines	Mine											✓
Siltstone	Coedydd Dyffryn Ffestiniog (Gogleddol)	Mine										✓	
Siltstone	Penygarnedd Mine	Mine										✓	
Slate	Aire Point to Carrick Du	Mine					✓						✓
Slate	Bulkamore Iron Mine	Mine							✓			✓	✓
Slate	West Llangynog Slate Mine	Mine					✓		✓		✓	✓	✓
Tuff	Glynllifon	Mine					✓		✓			✓	
Tuff	Mwyngloddiau a Chreigiau Gwydyr	Mine					✓		✓		✓	✓	

Table A4 Railway tunnels that are notified as SSSI in respect of bat species.

ROCK	SSSI	TYPE	B. bar	M. myo	M. bec	M. bra	M. dau	M. mys	M. nat	P. aur	R. hip	R. fer
							BAT SPECIES					
Chalk	Singleton and Cocking Tunnels	Railway tunnel	✓	✓	✓	✓	✓	✓	✓	✓		
Chalk	Buckshraft Mine & Bradley Hill Railway Tunnel	Railway tunnel										✓
Diorite & tonalite	The Malvern Hills	Railway tunnel									✓	
Limestone	Buckshraft Mine & Bradley Hill Railway Tunnel	Railway tunnel and mine										✓

Eighty-five popular climbing crags, the counties they occur in and the rocks they are made of

CRAG	COUNTY	ROCK
Almscliffe	Yorkshire	Gritstone
Aonach Dubh	Argyll & Bute	Andesite
Avon Gorge	Avon	Limestone
Ben Nevis	Highlands	Andesite
Bidean nam Bian	Highlands	Rhyolite
Beinn Eighe	Highlands	Quartzite
Berry Head	Devon	Limestone
Binnean Shuas	Highlands	Granite
Birchen Edge	Derbyshire	Gritstone
Bla Bheinn	Isle of Skye	Gabbro
Black Crag	Cumbria	Gabbro
Black Cuillin Ridge	Isle of Skye	Gabbro/Basalt
Bosigran	Cornwall	Granite
Bowfell	Cumbria	Rhyolite
Buachaille Etive Mòr	Highlands	Rhyolite
Carnmore Crag	Highlands	Gneiss
Carreg Wastad	Gwynedd	Rhyolite
Castel Cidwm	Gwynedd	Rhyolite
Chair Ladder	Cornwall	Granite
Cheddar Gorge	Somerset	Limestone
Chee Dale	Derbyshire	Limestone
Cir Mhor	Arran	Granite
Clogwyn Du'r Arddu	Gwynedd	Rhyolite
Clogwyn y Grochan	Gwynedd	Rhyolite
Cobbler, The	Argyll & Bute	Schist
Coire an Lochain	Invernesshire	Granite
Craig Cywarch	Gwynedd	?
Craig Gogarth	Anglesey	Quartzite

CRAG	COUNTY	ROCK
Craig y Castell	Gwynedd	Dolerite
Craig yr Ysfa	Gwynedd	Rhyolite
Cratcliffe Tor	Derbyshire	Gritstone
Creagan a'Choire Etchachan	Aberdeenshire	Granite
Creag a' Bhancair	Argyll & Bute	Rhyolite
Creag an Dubh Loch	Sutherland	Gneiss
Curbar Edge	Derbyshire	Gritstone
Cyrn Las	Gwynedd	Rhyolite
Dewerstone	Devon	Granite
Dinas Cromlech	Gwynedd	Rhyolite
Dinas Mot	Gwynedd	Rhyolite
Dove Crag	Cumbria	Rhyolite
Dow Crag	Cumbria	Rhyolite
Esk Buttress	Cumbria	Rhyolite
Froggatt Edge	Derbyshire	Gritstone
Gable Crag	Cumbria	Rhyolite
Garbh Choire	Aberdeenshire	Granite
Gillercombe	Rhyolite	Cumbria
Gimmer Crag	Rhyolite	Cumbria
Glen Coe	Highlands	Rhyolite
Glyder Fach	Gwynedd	Rhyolite
Glyder Fawr	Gwynedd	Rhyolite
Goat Crag	Cumbria	Rhyolite
Great Gable	Cumbria	Rhyolite
Hell's Lum	Moray	Granite
Hen Cloud	Staffordshire	Gritstone
Heron Crag	Cumbria	Rhyolite
High Tor	Derbyshire	Limestone
Llech Ddu	Gwynedd	Rhyolite
Lliwedd	Gwynedd	Rhyolite
Lochnagar	Aberdeenshire	Granite
Lundy	Devon	Granite
Malham Cove	North Yorkshire	Limestone
Milestone Buttress	Gwynedd	Rhyolite
Mowingword	Pembrokeshire	Limestone
Mother Carey's Kitchen	Pembrokeshire	Limestone
Old Man of Hoy, The	Orkney	Sandstone
Pabbay	Hebrides	Gneiss
Penyghent	Yorkshire	Gritstone
Pillar Rock	Cumbria	Rhyolite
Rannoch Wall	Highlands	Rhyolite

CRAG	COUNTY	ROCK
Raven Crag	Cumbria	Rhyolite
Roaches, The	Staffordshire	Gritstone
Scafell Pinnacle	Cumbria	Rhyolite
Sennen	Cornwall	Granite
Shelterstone Crag	Moray	Granite
Sgurr a' Chaorachain	Highlands	Sandstone
Shepherd's Crag	Cumbria	Rhyolite
South Stack	Anglesey	Quartzite
Sron na Ciche	Isle of Skye	Gabbro
Stackpole Head	Pembrokeshire	Limestone
Stanage Edge	Derbyshire/Yorkshire	Gritstone
Stoney Middleton	Derbyshire	Limestone
Swanage	Dorset	Limestone
Tremadog	Gwynedd	Dolerite
Trilleachan Slabs	Argyll & Bute	Granite
Tryfan	Gwynedd	Rhyolite

Notes

Chapter 1   Rationale

1. The English language is a wonderful tool that is gradually being damaged by misuse: 'ancient' in the context of woodland is an excellent example.

    Correctly applied, the word 'ancient' denotes something belonging to the very distant past and no longer in existence. It is an anthropological term that is typically applied to civilisations and their beliefs.

    Therefore, if it was applied to vegetation it would mean long extinct. To give an example, *Araucarioxylon arizonicum* is an ancient tree. Even supposing we were to accept that the species would not have to be extinct for an individual *Quercus robur* to be ancient, it would only warrant ancient status long after it had died and decayed to nothing. We might thus have a painting of a tree that was long gone and refer to that specific tree as ancient.

    Atlantis (if it ever existed) would be ancient and *Homo erectus* are ancient, but Stonehenge is not ancient and neither is the race that built it. It might, however, be reasonable to suggest that the religion for which it was built is ancient.

2. Niche Partitioning theory is an attempt to explain how several different species of the same subfamily can exploit the same logistical resource environment, and yet coexist without interspecies conflict.

    The niche dimensions encompass multiple macro-, meso- and micro-environmental resources, and an organism can only exist in a given situation if a subset of all the resources offered are within the limits to which it is adapted. These different resources can be plotted on different axes, or 'environmental gradients', and the 'niche' is the point where all those gradients intersect.

    Different subfamilies all exist on the same gradient, but occupy different niches within which they have a competitive superiority. This is important. Competition for resources is a race for a particular resource in space and time, and a superiority in defence once won. Unless both species had the ability to exploit the resource there would not have been competition. We do not see situations of access and exclusion: we see superiority and inferiority in time and/or space.

    In order to coexist without conflict, we see stratification of the resource environment. The niche partitions of all species have been defined over millennia by competition for resources, and they have evolved to occupy stratified positions of each gradient and thereby avoid open war.

3. See https://www.fscbiodiversity.uk/blog/what-biological-record and the rest of this chapter.

4. Semantic memory refers to a portion of long-term memory that processes ideas and concepts that are not drawn from personal experience.

Chapter 2   An introduction to rock

5. Whether and for how long each day they are open to the sun.
6. Which influences humidity.
7. Mudstone and siltstone are typically interlaminated.
8. Whether the information will have any particular value is unknown but it seemed sensible to collate it while the review was being performed in case any particular pattern was discernible as the bat data set grew.

9. Long after you are dead, by which time you will have discovered and recorded so many new rock face roosts that you will be so famous that someone else will want to carry on your work.
10. https://mapapps.bgs.ac.uk/geologyofbritain/home.html
11. See Chapter 2.
12. See Chapter 3.
13. Grey long-eared bats *Plecotus austriacus* are crevice-roosting species and they are grey. Rock is grey … is this pelage the result of natural selection that favours grey pelage that is background-matched to a grey substrate? Is the grey long-eared bat a species of granite tors and limestone escarpments?
14. Currently from: https://hub.jncc.gov.uk/assets/9578d07b-e018-4c66-9c1b-47110f14df2a
15. Those of you in consultancy may begin to find yourself actually enjoying your career rather than looking like some dead-eyed Gelphling with all the 'life essence' sucked out of them. (All graduates start their careers either like Tigger, Roo or Piglet; some only ever progress to being rather Pooh, and most finish as either Eeyore or worse: Owl. (Hopefully, you will take a more objective approach that will lead to a fulfilling career.)

Chapter 4   Loose rock: characterising and recording the landforms and the Potential Roost Features they may hold

16. A destructive search comprises the breaking apart of the habitat in order to find the organism occupying it. In the context of loose rock, lifting rocks would be a destructive search if they could not be put back exactly as they had been.
    Destructive searches carry the risk that the organism may be injured or killed in the process. This risk will obviously be increased if the person performing the search does not know where the organism is, and increased still further if the organism is mobile inside the habitat.
17. https://mapapps.bgs.ac.uk/geologyofbritain/home.html
18. Concrete and brick rubble dumped on brownfield sites are not considered because they tend to be piles and do not have sufficiently similar topographically to be considered artificial scree of blockfield. This does not mean that bats do not use dumped rubble, it just means that it is not part of this project. Gabion baskets are not considered either, but if they are not used by roosting whiskered bats it would be very surprising.
19. Currently from: https://hub.jncc.gov.uk/assets/9578d07b-e018-4c66-9c1b-47110f14df2a
20. The Phase 1 handbook places gorse in scrub, and suggests that scrub may be a climax community: this is wrong. Simms (1971) correctly and accurately described scrub as regenerating woodland composed of bushes and trees no taller than 8 m. Gorse stands are a habitat in their own right.

Chapter 5   Subterranean rock: characterising and recording the landforms and the Potential Roost Features they may hold

21. Associated with limestone around Derbyshire (British Geological Survey website https://www.bgs.ac.uk/data/publications/pubs.cfc?method=viewRecord&publnId=19867792).
22. Karst is a topography formed from the dissolution of soluble rocks.
23. It seems unlikely that none of the sea caves below the Pembroke Ranges extend back further than 100 m and it is astonishing that the west coast of the Republic of Ireland has no documented long sea caves. It may be that their omission on the list is due to a lack of recording rather than a lack of length.
24. A bell-pit is the most primitive mining method and common to prehistoric times. It comprises a vertical shaft which follows a seam, and is then enlarged in the base so that the void is bell or flask-shaped (Hayman 2016). The pits were typically a maximum 5 m deep and as they could only be worked out a short way in the base before the roof collapsed, they were dug in lines along the seam (Hayman 2016).
25. The longwall method of mining allows the entirety of a horizontal seam of coal to be removed with the roof being held up by wooden pit props and worthless rock and slack.
26. Drive and fan is a method of extracting ball clay where a shaft is sunk, then adits ('drives') were dug in opposite directions from the floor following the level 'strike' of the clay seam.

These faces of the drives were then worked round in successive ranks, pivoting from the base of the shaft like a fan, until both drives met (see Edwards (2011) for a detailed explanation with figures).

27. Pillar and stall is a method of extracting limestone (Edwards 2011), ochre (Clarke *et al.* 2012) and coal (Hayman 2016). A broadly horizontal adit is gradually widened, with the roof supported by leaving behind pillars of unworked stone (Edwards 2011). The method was the most common technique for extracting coal in deep mines before 1800 but was superseded by the longwall method (Hayman 2016).

28. The maximum reach from a shaft that I could find is listed by Burr (2015a) as 91 m.

29. Canal tunnels are specifically not included in the BRHK project because so many do not have towpaths and there was insufficient time to begin investigating them.

30. In some accounts of subterranean roosting, smaller 'hidden' PRF are referred to as 'cryptic'. The word 'cryptic' is not correct when describing PRF. It is correctly used in consideration in the pelage and morphology of the bats themselves. In this account, the word 'enveloping' has been adopted to encompass all situations where the bat is in a confined space, that would by virtue of its physical characteristics be less visible. The cryptic coloration visible in the countershading of bat some species, and the potential background matching function suggested by the pelage and morphology of others, is a story for another day.

31. Significant changes would include new entrances being opened or an old one closed, and a system that had been wet gradually drying, or the opposite.

32. There is already an excellent account provided by Zukal, J., Berková, H., Banďouchová, H., Kováčová, V. & Pilula, J. 2017. Bats in caves: activity and ecology of bats wintering in caves. *Intech.* http://dx.doi.org/10.5772/intechopen.69267

33. The more observant of you will note that the greater proportion of the reference material used in this book is the result of studies carried out in countries far removed from the British Isles. However, in 2021 it represents the best available evidence. Notwithstanding, it should be borne in mind that although the studies relate to species that are native here, the differences between our climate and that of the country in which the accounts were recorded may be significant. As a result, the accounts may not be fully accurate representations of the behaviour of the same species in the British Isles.

34. Cave maps often name specific features. For example, Goatchurch Cavern in the Mendips (in which many young cavers first get a taste for spelunking) encompasses: giant's stairs, the coffin lid, bloody tight, the boulder maze, the coal chute, and the drain pipe.

35. Double recessed features are a common feature of all the subterranean landforms. In solution caves we find domed cupolas with integral ceiling pockets (if you were making a mould for a pair of novelty breasts, the bat would be occupying the nipple). In mines we find manholes with wall pockets in the bare rock, and in railway tunnels we see manholes with drain-holes offering ceiling pockets left for drainage in the brickwork lining.

36. https://mapapps.bgs.ac.uk/geologyofbritain/home.html

37. Notable accounts are those given by Daan & Wichers (1968) and Kokurewicz (2004), but these are in specific situations on the Continent.

38. Or in Blancmange (the band, not the dessert).

39. A bar of rock extending from the floor to the ceiling and formed by a stalactite and stalagmite fusing. Unlike a pillar, a column is formed following the phreatic stage of the cave level when the system is above the water table.

40. A column of rock remaining after solution of the rock on either side or deliberately left in place to support a mine ceiling. Unlike a column, a pillar is formed during the phreatic stage of the cave level when the system was full of water.

41. Residual rock spanning a passage, typically the result of water erosion. Bridges may be narrow and simply have apertures between the bridge and the floor below and ceiling above, or they may continue for a distance with a passage below and a bridge-tube above.

42. A bridge-tube is a typically low-diameter squeeze throughway created above a passage by a bridge.

43. Clearly this will have a marked bearing on attempts to get lesser horseshoe bats to adopt railway tunnels.

44. Ecochaeology is the practice of performing commissioned surveillance to test what is at best a subjective hypothesis (if there is any hypothesis at all). Ecochaeology is never paid for by the person who demands it, nor will it be put to any practical use by the person that demands it, and nor will the person that demands it collate the data, store the data, or learn anything from the data.

      Years from now archaeologists will look back at the practice and pigeon-hole it with that old chestnut, 'it will have been of ritual significance'. This is not all that far from the truth, because it is a part of the cult of ecology where the people making the demands are accorded 'Papal Infallibility' that puts their rantings above science and planning guidance, and even professional courtesy.
45. Although exactly what defines 'a relatively low humidity' is not defined, so the information is functionally useless.

Chapter 6   Advice for anyone proposing to survey a rock landform or monitor an artificial landform to deliver a new roost for a specific bat species

46. And potentially the greater mouse-eared bat (if it returns to our shores) and Alcathoe's bat (if we later find out they use caves).
47. In order to allow data gathering in support of Impact Assessments and for research, licences are made available from Natural England (NE), Natural Resources Wales (NRW), Department of Agriculture, Environment and Rural Affairs (DAERA) and Scottish Natural Heritage (SNH). To qualify for such a licence the surveyor must have demonstrated a robust understanding of bat ecology and competence in the methods covered by the licence.
48. It is hoped that: **a)** the unlicensed naturalist will support their approach with a completed recording form; and, **b)** anyone approached to assist will be encouraging and helpful to the unlicensed naturalist. I myself will assist anyone in south-west England and south Wales, but I will need to see the recording form as evidence that the request is not part of a posing/fondling ego-trip.
49. A great deal of roost searching and photography that is done under the guise of science looks suspiciously like posing and fondling. The recording form is more important than the endoscope. If you are not collecting data that has a value to the bats then you should not be disturbing them. If you are not creating a map you do not need a camera. And if you are only interested in seeing and having your photo taken with small furry creatures, buy a guinea pig and stay at home. Rant not over … *to be continued until the day I die …*
50. Only half the blame for this lies with the consultants who trousered the cash, the other half lies with the conservation agencies who have just let it run.
51. Measurable as a quantitative value using a standard scale which is recorded using tools to which everyone has access. Where a qualitative scale is used, it should wherever possible be a dichotomy (e.g. yes/no; alive and healthy/dead or unhealthy; present/absent; etc.) which is achieved using a simple set of criteria that cannot be applied subjectively.
52. Shed felt is really useful for testing the effect of partitions because it is cheap as a temporary installation, and you can cut it back to test the response of a reduction in surface area. When you finally get the conditions right it can be replaced with a permanent partition using the felt as a pattern to get the size and shape right first time.

References

Alberdi, A., Aihartza, J., Aizpurua, O., Salsamendi, E., Brigham, R.M. & Garin, I. 2014. Living above the treeline: roosting ecology of the alpine bat *Plecotus macrobullaris*. *European Journal of Wildlife Research*, 61: 17–25. https://doi.org/10.1007/s10344-014-0862-8

Altringham, J. 2003. *British Bats*. HarperCollins New Naturalist Series, London.

Ancillotto, L., Cistrone, L., Mosconi, F., Jones, G., Boitani, L. & Russo, D. 2014. The importance of non-forest landscapes for the conservation of forest bats: lessons from barbastelles (*Barbastella barbastellus*). *Biodiversity and Conservation* 24(1): 171–185. https://doi.org/10.1007/s10531-014-0802-7

Antikainen, E. 1978. The breeding adaptation of the Jackdaw *Corvus monedula* L. in Finland. *Savonia* 2: 1–45.

Balch, H. 1937. *Mendip, its Swallet Caves and Rock Shelters*. Cathedral Press, Wells.

Barrett-Hamilton, G. 1910. *A History of British Mammals: Vol. 1 – Bats*. Gurney & Jackson, London. https://doi.org/10.5962/bhl.title.55827

Bels, L. 1952. Fifteen years of bat banding in the Netherlands. *Publ. natuurhist. Gen. Limberg* 5(1): 1–99.

Bezem, J., Sluiter, J. & van Heerdt, P. 1960. Population statistics of five species of the bat genus *Myotis* and one of the genus *Rhinolophus*, hibernating in the caves of S. Limburg. *Arch. Neerl. De Zool.* 13(4): 511–539. https://doi.org/10.1163/036551660X00170

Bezem, J., Sluiter, J. & van Heerdt, P. 1964. Some characteristics of the hibernating locations of various species of bats in south Limburg. *Proceedings of the Koninklijke Nederlandse Akademie van Wettenschappen, Amsterdam* 67: 325–350.

Billington, G. 2000. *Holnicote Estate – Horner Woods Bat Survey – Somerset*. Greena Ecological Consultancy, Frome.

Billington, G. 2004. *Determination of Autumn and Winter Use of South Pembrokeshire Cliffs by Horseshoe Bats*. Contract Science Report 619. Countryside Council for Wales, Bangor.

Bogdanowicz, W. 1983. Community structure and interspecific interactions in bats hibernating in Poznań. *Acta Theriologica* 28(23): 357–370. https://doi.org/10.4098/AT.arch.83-31

Bogdanowicz, W. & Urbańczyk, Z. 1983. Some ecological aspects of bats hibernating in city of Poznań. *Acta Theriologica* 28(24): 371–385. https://doi.org/10.4098/AT.arch.83-32

Bright, P., Morris, P. & Mitchell-Jones, T. 2006. *The dormouse conservation handbook* (2nd edn). English Nature, Peterborough.

BTHK 2018. *Bat Roosts in Trees: A Guide to Identification and Assessment for Tree-Care and Ecology Professionals*. Pelagic Publishing, Exeter.

Bunnell, D. 2004. Littoral caves. In: Gunn, J. (ed.). 2004. *Encyclopedia of Caves and Karst Science*. Taylor & Francis, London. https://doi.org/10.4324/9780203483855

Burr, P. 2015a. *Mines and Minerals of the Mendip Hills, Volume 1*. Mendip Cave Registry and Archive, Somerset.

Burr, P. 2015b. *Mines and Minerals of the Mendip Hills, Volume 2*. Mendip Cave Registry and Archive, Somerset.

Chamberlain, A. 2004. Britain and Ireland: archaeological and paleontological caves. In: Gunn, J. (ed.). 2004. *Encyclopedia of Caves and Karst Science*. Taylor & Francis, London. https://doi.org/10.4324/9780203483855

Chanin, P. & Woods, M. 2003. Surveying dormice using nest tubes: results and experiences from the South West Dormouse Project. English Nature Research Report 524. English Nature, Peterborough.

Chapman, P. 1993. *Caves and Cave Life*. HarperCollins New Naturalist Series, London.

Clarke, M., Gregory, N. & Gray, A. 2012. *Earth Colours: Mendip and Bristol Ochre Mining*. Mendip Cave Registry and Archive, Somerset.

Cottle, R. & John, S. 2009. Coastal ecology and geomorphology. In: Morris, P. & Therivel, R. (eds). 2009. *Methods of Environmental Impact Assessment*, 3rd edn. Routledge Taylor & Francis, London & New York.

Coward, T. 1906. On some habits of the lesser horseshoe bat *Rhinolophus hipposideros*. *Proceedings of the Zoological Society* 76 (3–4): 849–855.

Coward, T. 1907. On the winter habits of the Greater Horseshoe Rhinolophus ferrumequinum (Schreber), and other cave haunting bats. *Proceedings of the Zoological Society of London* 77(2): 312–324.

Crammer, H. 1899. Eishöhlen-und Windröhren Studien. *Vienna Geographischen Gesellschaft* 1: 19–76.

Cropley, J. 1965. Influence of surface conditions on temperatures in large cave systems. *Bull. Nat. Speleol. Soc.* 27: 1–10.

Cullingford, C. (ed.). 1962. *British Caving: An Introduction to Speleology*, 2nd edn. Routledge & Kegan Paul, London.

Daan, S. 1973. Activity during natural hibernation in three species of Vespertilionid bats. *Netherlands Journal of Zoology* 23(1): 1–71. https://doi.org/10.1163/002829673X00193

Daan, S. & Wichers, H. 1968. Habitat selection of bats hibernating in a limestone cave. *Z. Säugetierk* 33: 262–287.

de Freitas, C. & Littlejohn, R. 1987. Cave climate: assessment of heat and moisture exchange. *J. Climatol.* 7: 553–569. https://doi.org/10.1002/joc.3370070604

de Freitas, C., Littlejohn, R., Clarkson, T. & Kristament, L. 1982. Cave climate: assessment of airflow and ventilation. *Int. J. Climatol.* 2: 383–397. https://doi.org/10.1002/joc.3370020408

Degn, H.J. 1989. Summer activity of bats at a large hibernaculum. In: Hanák, V., Horáček, I. & Gaisler, J. (eds). 1989. *European Bat Research 1987*, pp. 523–526. Charles Univ. Press, Praha.

Dietz, C. & Kiefer, A. 2016. *Bats of Britain and Europe*. Bloomsbury, London.

Dietz, C., von Helversen, O. & Nill, D. 2011. *Bats of Britain, Europe and Northwest Africa*. A. & C. Black, London.

Dyer, S. 2013. Sam Dyer Ecology and Gwynedd Bat Group: The North Wales Serotine Project. Talk at the Bat Conservation Trust 2013 Wales Conference.

Edwards, R. 2011. *Devon's Non-Metal Mines: Discovering Devon's Slate, Culm, Whetstone, Beer Stone, Ball Clay and Lignite Mines*. Halsgrove, Somerset.

Farrant, A. & Harrison, T. 2017. Hypogenic caves in the UK. In: Klimchouk, A., Palmer, A., De Waele, J., Auler, A. & Audra, P. (eds). 2017. *Hypogene Karst Regions and Caves of the World*, pp. 43–60. Springer International. https://doi.org/10.1007/978-3-319-53348-3_2

Fitter, R. & Fitter, F. 1967. *The Penguin Dictionary of British Natural History*. Penguin, Middlesex.

Folk, C. 1968. Das nisten and die populationsdynamic der Dohle (*Corvus monedula* L.) in der CSSR. *Zoologické Listy* 17: 221–236.

Francou, B. & Manté, C. 1990. Analysis of the segmentation in the profile of alpine talus slopes. *Permafrost and Periglacial Processes* 1: 53–60. https://doi.org/10.1002/ppp.3430010107

Gaisler, J. 1970. Remarks on the thermopreferendum of palearctic bats in their natural habitats. *Bijdragen Tot De Dierkunde* 40(1): 33–35. https://doi.org/10.1163/26660644-04001010

Gaisler, J. & Chytil, J. 2002. Mark-recapture results and changes in bat abundance at the cave of Na Turdoldu, Czech Republic. *Folia Zool.* 51(1): 1–10.

Gallon, R. 2020. *Trogloneta granulum* Simon, 1922 in North Wales (Araneae, Mysmenidae). *British Arachnological Society Newsletter* 147: 15–16.

Garlick, S. 2009. *Flakes, Jugs, and Splitters: A Rock Climber's Guide to Geology (How To Climb Series)*. FalconGuides, Richmond.

Gilbert, O. 2000. *Lichens*. Collins New Naturalist Series, London.

Gleick, J. 1987. Sometimes heavier objects go to the top: here's why. *The New York Times*, 24 March. https://www.nytimes.com/1987/03/24/science/sometimes-heavier-objects-go-to-the-top-here-s-why.html

Glover, A. & Altringham, J. 2008. Cave selection and use by swarming bat species. *Biological Conservation* 141: 1493–1504. https://doi.org/10.1016/j.biocon.2008.03.012

Halliday, W. 2004. Talus caves. In: Gunn, J. (ed.). 2004. *Encyclopedia of Caves and Karst Science*. Taylor & Francis, London. https://doi.org/10.4324/9780203483855

Harmata, W. 1969. The thermopreferendum of some species of bats (Chiroptera). *Acta Theriologica* 14(5): 49–62. https://doi.org/10.4098/AT.arch.69-5

Harmata, W. 1987. The frequency of winter sleep interruptions in two species of bats hibernating in limestone tunnels. *Acta Theriologica* 32(21): 331–332. https://doi.org/10.4098/AT.arch.87-23

Harvey, P., Nellist, D. & Telfer, M. (eds). 2002a. *Provisional Atlas of British Spiders (Arachnida, Araneae), Volume 1*. Biological Records Centre, Huntingdon.

Harvey, P., Nellist, D. & Telfer, M. (eds). 2002b. *Provisional Atlas of British Spiders (Arachnida, Araneae), Volume 2*. Biological Records Centre, Huntingdon.

Hayman, R. 2016. *Coal Mining in Britain*. Shire, Oxford.

Hellawell, J. 1991. Development of a rationale for monitoring. In: Goldsmith, F. (ed.). 1991. *Monitoring for Conservation and Ecology*, pp. 1–14. Chapman & Hall, London. https://doi.org/10.1007/978-94-011-3086-8_1

Hill, D., Fasham, M., Tucker, G., Shewry, M. & Shaw, P. (eds). 2005. *Handbook of Biodiversity Methods: Survey, Evaluation and Monitoring*. Cambridge University Press, Cambridge. https://doi.org/10.1017/CBO9780511542084

Holyoak, D. 2009. Breeding biology of the Corvidae. *Bird Study* 14(3): 153–168. https://doi.org/10.1080/00063656709476159

Hooper, J. & Hooper, W. 1956. Habits and movements of cave-dwelling bats in Devonshire. *Proc. Zool. Soc. Lond.* 127: 1–26. https://doi.org/10.1111/j.1096-3642.1956.tb00457.x

Jones, G., Duverge, P. & Ransome, R. 1995. Conservation biology of an endangered species: field studies of greater horseshoe bats. *Zoological Society of London Symposia* 67: 309–324.

Jones, I. 2003. *Victorian Slate Mining: A Social and Economic Study*. Landmark, Derbyshire.

JNCC 1998. *Common Standards for Monitoring Designated Sites: A Statement on Common Standards Monitoring*. Joint Nature Conservation Committee, Peterborough.

JNCC 2010. *Handbook for Phase 1 Habitat Survey: A Technique for Environmental Audit*. Joint Nature Conservation Committee, Peterborough.

Judson, D. (ed.). 1984. *Caving Practice & Equipment*. David & Charles, London.

Kearey, P. 1996. *The New Penguin Dictionary of Geology*. Penguin, London.

Kerney, M. 1999. *Atlas of the Land and Freshwater Molluscs of Britain and Ireland*. Harley, Colchester.

Kleinman, D., Reading, P., Miller, B., Clark, T., Scott, M., Robinson, J. & Wallace, R. 2000. Improving the evaluation of conservation programs. *Conservation Biology* 14(2): 356–365. https://doi.org/10.1046/j.1523-1739.2000.98553.x

Klimchouk, A. 2000. Speleogenesis under deep-seated and confined settings. In: Klimchouk, A., Ford, D., Palmer, A. & Dreybrodt, W. (eds). 2000. *Speleogenesis: Evolution of Karst Aquifers*, pp. 244–260. National Speleological Society, Huntsville.

Klimchouk, A. 2004. Caves. In: Gunn, J. (ed.). 2004. *Encyclopedia of Caves and Karst Science*. Taylor & Francis, London. https://doi.org/10.4324/9780203483855

Klys, G. 2013. Effect of microclimate of underground systems on the occurance of hibernating bats. *Journal of Environmental Science and Engineering* B(2): 36–45.

Kokurewicz, T. 2004. Sex and age related habitat selection and mass dynamics of Daubenton's bats *Myotis daubentonii* (Kuhl, 1817) hibernating in natural condition. *Acta Chiropterol* 6: 121–144. https://doi.org/10.3161/001.006.0110

Kowalski, K. 1953. Materialy do rozmieszczenia i ekologii nietoperzy jaskiniowych v.' Polsce. *Fragm. faun. Mus. zool. Pol.* 6(21): 541–567. In: Daan, S. & Wichers, H. 1968. Habitat selection of bats hibernating in a limestone cave. *Z. Säugetierk* 33: 262–287. https://doi.org/10.3161/15053970FF1949.6.21.541

Kristiansen, K. 2018. Hide and seek: a pilot study on day roosts in autumn and hibernacula for Vesper bats in Southeast Norway. Master's thesis, Norwegian University of Life Sciences.

Legg, C. & Nagy, L. 2006. Why most conservation monitoring is, but need not be, a waste of time. *Journal of Environmental Management* 78: 194–199. https://doi.org/10.1016/j.jenvman.2005.04.016

Lesiński, G. 1986. Ecology of bats hibernating underground in central Poland. *Acta Theriologica* 31(37): 507–521. https://doi.org/10.4098/AT.arch.86-45

Lesiński, G. 1989. Summer and autumn dynamics of *Myotis daubentonii* in underground shelters in central Poland. In: Hanák, V., Horáček, I. & Gaisler, J. (eds). 1989. *European Bat Research 1987*, pp. 519–521. Charles Univ. Press, Praha.

Littva, J., Bella, P., Gaál, Ľ., Holúbek, P. & Hók, J. 2017. Extraordinary geology and fault-controlled phreatic origin of the Zápoľná Cave (Kozie chrbty Mountains, Slovakia). *Acta Geologica Slovaca* 9(1): 25–34.

Luckman, B. 2013. Mountain and hillslope geomorphology 7.17: Processes, Transport, Deposition, and Landforms: Rockfall. In: Schroder, J. (ed.). 2013. *Treatise on Geomorphology*, pp. 174–182. Elsevier, London.

Mackay, J. & Burrous, C. 1979. Uplift of objects by an upfreezing ice surface. *Can. Geotech. J.* 16: 609–613. https://doi.org/10.1139/t79-065

Margoluis, R. & Salafsky, N. 1998. *Measures of Success: Designing, Managing, and Monitoring Conservation and Development Projects*. Island Press, Washington.

Mason, C. & Lyczynski, F. 1980. Breeding biology of the Pied and Yellow Wagtails. *Bird Study* 27(1): 1–10. https://doi.org/10.1080/00063658009476650

McKay, R. 1996. Conservation and ecology of the red-billed chough *Pyrrhocorax pyrrhocorax*. PhD thesis, University of Glasgow.

Michaelsen, T. & Grimstad, K. 2008. Rock scree: a new habitat for bats. *Nyctalus* 13: 122–126.

Michaelsen, T., Olsen, O. & Grimstad, K. 2013. Roosts used by bats in late autumn and winter at northern latitudes in Norway. *Folia Zool.* 62(4): 297–303. https://doi.org/10.25225/fozo.v62.i4.a7.2013

Mihál, T. & Kaňuch, P. 2006. Habitat factors influencing bat assemblages hibernating in abandoned mines in the Štiavnické vrchy Mts (Slovakia): preliminary results. *Nyctalus* 11(4): 293–301.

Mihevc, A., Slabe, T. & Šebela, S. 2004. Morphology of caves. In: Gunn, J. (ed.). 2004. *Encyclopedia of Caves and Karst Science.* Taylor & Francis, London. https://doi.org/10.4324/9780203483855

Mikula, P. 2013. Western Jackdaw (*Corvus monedula*) attacking bats (*Chiroptera*): observations from Bardejov, northeastern Slovakia. *Silvia* 49: 157–159.

Mitchell, D. 2008. *Core Management Plan for Tanat and Vyrnwy Bat Sites Special Area of Conservation.* Countryside Council for Wales, Bangor.

Mitchell-Jones, A. & McLeish, A. 2004. *Bat Workers' Manual,* 3rd edn. Joint Nature Conservation Committee, Peterborough.

Moore, D. 1954. Origin and development of sea caves. *National Speleological Society Bulletin* 16: 71–76.

Murariu, D. & Gheorghiu, V. 2010. Şura Mara Cave (Romania), the most important known hibernating roost for *Pipistrellus pygmaeus* Leach, 1825 (Chiroptera: Vespertilionidae). *Travaux du Muséum National d'Histoire Naturelle* 53: 329–338. https://doi.org/10.2478/v10191-010-0023-6

Mylroie, J. & Carew, J. 1987. Field evidence of the minimum time for speleogenesis. *National Speleological Society Bulletin* 49: 67–72.

Nagy, Z. & Postawa, T. 2010. Seasonal and geographical distribution of cave-dwelling bats in Romania: implications for conservation. *Anim. Conserv.* 14: 74–86. https://doi.org/10.1111/j.1469-1795.2010.00392.x

National Museums Liverpool. 2018. *The Status and Distribution of the Ground Beetle* Leistus montanus *in 2017 at its Historic Sites in Wales.* NRW Evidence Report No. 249. National Resources Wales, Bangor.

Natural England. nd. SSSI detail – Natural England: Designated Sites/Beer Quarry & Caves. https://designatedsites.naturalengland.org.uk

Palmer, A. & Audra, P. 2004. Patterns of caves. In: Gunn, J. (ed.). 2004. *Encyclopedia of Caves and Karst Science.* Taylor & Francis, London. https://doi.org/10.4324/9780203483855

Parker, N., Mackay, C. & Webb, K. 2013. *Glyn Rhonwy Pumped Storage Development Consent Order. Appendix 7.15 AECOM (2013d) Glyn Rohnwy Pumped Storage Scheme – Bat Mitigation Report April 2013.* Report prepared by AECOM for Snowdonia Pumped Hydro.

Parnell, I. 2020. *Hard Rock: Great British Rock Climbs from VS to E4.* Vertebrate, Sheffield.

Parry, S. 2011. Developments in earth surface processes. In: Smith, M., Paron, P. & Griffiths, J. (eds). 2011. *Geomorphological Mapping: Methods and Applications: Volume 15,* 413–440. Elsevier, London.

Perry, R. 2013. A review of factors affecting cave climates for hibernating bats in temperate North America. *Environ. Rev.* 21: 28–39. https://doi.org/10.1139/er-2012-0042

Peters, R. 1991. *A Critique for Ecology.* Cambridge University Press, Cambridge.

Piksa, K. & Nowak, J. 2013. The bat fauna hibernating in the caves of the Polish Tatra Mountains, and its long-term changes. *Cent. Eur. Biol.* 8(5): 448–460. https://doi.org/10.2478/s11535-013-0146-9

Pill, A. 1951. Jug Holes caves and its bats. *Naturalist* 836: 7.

Poulson, T. & White, W. 1969. The cave environment: limestone caves provide unique natural laboratories for studying biological and geological process. *Science* 165(3897): 971–981. https://doi.org/10.1126/science.165.3897.971

Poulton, S. 2006. *An Analysis of the Usage of Batboxes in England, Wales & Ireland for The Vincent Wildlife Trust.* Vincent Wildlife Trust, Ledbury.

Pragnell, H. 2016. *Early British Railway Tunnels: The Implications for Planners, Landowners and Passengers Between 1830 and 1870.* University of York Railway Studies, York.

Punt, A. & van Nieuwenhoven, P. 1957. The use of radioactive bands in tracing hibernating bats. *Experientia, Basel* 13(1): 51–54. https://doi.org/10.1007/BF02156962

Ransome, R. 1968. The distribution of the Greater horse-shoe bat, *Rhinolophus ferrumequinum*, during hibernation, in relation to environmental factors. *J. Zool. Lond.* 154: 77–112. https://doi.org/10.1111/j.1469-7998.1968.tb05040.x

Ransome, R. 1971. The effect of ambient temperature on the arousal frequency of the hibernating greater horseshoe bat, *Rhinolophus ferrumequinum*, in relation to site selection and hibernation state. *J. Zool. Lond.* 164: 353–371. https://doi.org/10.1111/j.1469-7998.1971.tb01323.x

Ransome, R. 1990. *The Natural History of Hibernating Bats.* Christopher Helm, London.

Reade, W. & Hosking, E. 1974. *Nesting Birds, Eggs and Fledglings.* Blandford, Poole.

Rixhon, G. & Demoulin, A. 2013. Glacial and periglacial geomorphology. In: Schroder, J. (ed.). 2013. *Treatise on Geomorphology*, volume 8, pp. 392–415. Elsevier, London.

Salafsky, N., Margoluis, R. & Redford, K. 2001. *Adaptive Management: A Tool for Conservation Practitioners.* Biodiversity Support Program, Washington.

Salafsky, N., Margoluis, R., Redford, K. & Robinson, J. 2002. Improving the practice of conservation: a conceptual framework and research agenda for conservation science. *Conservation Biology* 16: 1469–1479. https://doi.org/10.1046/j.1523-1739.2002.01232.x

Schober, W. & Grimmberger, E. 1993. *Bats of Britain and Europe.* Hamlyn, London.

Schober, W. & Grimmberger, E. 1997. *The Bats of Europe & North America.* T.F.H. Publications, Neptune City, NJ.

Schofield, H. 2008. *The Lesser Horseshoe Bat Conservation Handbook.* Vincent Wildlife Trust, Herefordshire.

Schofield, H. & Mitchell-Jones, A. 2003. *The Bats of Britain and Ireland.* Vincent Wildlife Trust, Herefordshire.

Sharrock, J. 1976. *The Atlas of Breeding Birds in Britain and Ireland.* T. & A.D. Poyser, London.

Shewring, M. 2021. Evidence of rock exposure roosting bat species in the UK from recreational rock climbers. Unpublished manuscript, Cardiff University.

Shiel, C., Jones, G. & Waters, D. 2008. Leisler's bat. In: Harris, S. & Yalden, D. (eds). 2008. *Mammals of the British Isles: Handbook*, 4th edn, pp. 334–338. Mammal Society, Southampton.

Simms, E. 1971. *Woodland Birds.* Collins New Naturalist Series, London.

Sluiter, J. & van Heerdt, P. 1964. Distribution and abundance of bats in S. Limburg from 1956 till 1962. *Ocerdruk Uit Het Natuurhistorisch Maandblad* 11(12): 164–173.

Sparrow, A. 2009. *The Complete Caving Manual.* Crowood Press, Wiltshire.

Stebbings, R. 1988. *The Conservation of European Bats.* Christopher Helm, London.

Tabor, J. 2011. *Blind Descent: The Quest to Discover the Deepest Place on Earth.* Constable & Robinson, London.

Tucker, G., Fasham, M., Hill, D., Shewry, M., Shaw, P. & Wade, M. 2005. Planning a programme. In: Hill, D., Fasham, M., Tucker, G., Shewry, M. & Shaw, P. (eds). 2005. *Handbook of Biodiversity Methods: Survey, Evaluation and Monitoring*, pp. 6–64. Cambridge University Press, Cambridge. https://doi.org/10.1017/CBO9780511542084.003

van Nieuwenhoven, P. 1956. Ecological observations in a hibernation quarter of cave-dwelling bats in South Limburg. *Publ. natuurhist. Gen. Limburg* 9(1): 1–56.

Vandel, A. 1965. *Biospeleology: The Biology of Cavernicolous Animals.* Pergamon, Oxford. https://doi.org/10.1016/B978-0-08-010242-9.50027-X

Vesey-Fitzgerald, B. 1949. *British Bats.* Methuen, London.

Vilborg, L. 1955. The uplift of stones by frost. *Geografiska Annaler* 37(3/4): 164–169. https://doi.org/10.1080/20014422.1955.11880874

Vos, P., Meelis, E. & Ter Keurs, W. 2000. A framework for the design of ecological monitoring programs as a tool for environmental and nature management. *Environmental Monitoring and Assessment* 61: 317–344. https://doi.org/10.1023/A:1006139412372

Waltham, A., Simms, M., Farrant, A. & Goldie, H. 1997. *Karst and Caves of Great Britain: Geological Conservation Review Series No. 12.* Chapman & Hall, London, pp. 819–842. https://doi.org/10.1017/S0016756898301508

Webb, P., Speakman, J., & Racey, P. 1996. How hot is a hibernaculum? A review of the temperatures at which bats hibernate. *Canadian Journal of Zoology* 74(4): 761–64. https://doi.org/10.1139/z96-087

Wermundsen, T. 2010. Bat habitat requirements – implications for land use planning. *Dissertationes Forestales* 111. https://doi.org/10.14214/df.111

Wermundsen, T. & Siivonen, Y. 2010. Seasonal variation in the use of winter roosts by five bat species in south-east Finland. *Cent. Eur. J. Biol.* 5: 262–273. https://doi.org/10.2478/s11535-009-0063-8

Whitaker, A. 1906. The flight of bats. *The Naturalist*: 379–384.

Whitten, D. & Brooks, J. 1979. *The Penguin Dictionary of Geology.* Penguin, London.

Wilson, J. 2012. Recreational caving. In: White, B. & Culver, D. (eds). 2012. *Encyclopaedia of Caves*, 2nd edn, pp. 641–648. Elsevier, London. https://doi.org/10.1016/B978-0-12-383832-2.00094-3

Wilson, K. 2007. *Classic Rock: Great British Rock Climbs.* Bâton Wicks, Sheffield.

Woodroffe, C. 2003. *Coasts: Form, Process and Evolution.* Cambridge University Press, Cambridge. https://doi.org/10.1017/CBO9781316036518

Yalden, D. & Morris, P. 1975. *The Lives of Bats.* David & Charles, Newton Abbot.

Zukal, J., Berková, H., Banďouchová, H., Kováčová, V. & Pilula, J. 2017. Bats in caves: activity and ecology of bats wintering in caves. *Intech.* http://dx.doi.org/10.5772/intechopen.69267

Index

References to figures and photographs appear in *italic* type; those in **bold** type refer to tables.